BARRON'S

Chemistry
Practice
Plus

400+ ONLINE QUESTIONS 🔍

WITH QUICK REVIEW GUIDE

MARK C. KERNION, M.A., AND JOSEPH A. MASCETTA, M.S., C.A.S.

Dedication

To

my wife, Patty; children, Emily, Sam, and Ben; granddaughters, Stella and Clare;
grandson, Nate; son-in-law, Rob; and daughter-in-law, Aly

—M. Kernion

To

my wife, Jean; daughters, Lisa, Linda, and Lori;
and their families,
who supported my efforts throughout the years

—J. Mascetta

Published by Kaplan North America, LLC, dba Barron's Educational Series

1515 W Cypress Creek Road
Fort Lauderdale, FL 33309
www.barronseduc.com

ISBN: 978-1-5062-8150-6

10 9 8 7 6 5 4 3 2 1

Kaplan North America, LLC, dba Barron's Educational Series print books are available at special quantity
discounts to use for sales promotions, employee premiums, or educational purposes. For more information or
to purchase books, please call the Simon & Schuster special sales department at 866-506-1949.

About the Authors

Mark Kernion, M.A.

Mark Kernion, M.A., taught AP, honors, and academic chemistry for 33 years at Mt. Lebanon High School in Pittsburgh, Pennsylvania, and currently teaches an online AP Chemistry course through PA Homeschoolers and Chemistry-Prep.com. He received the Yale Teaching Award in 2007, was inducted into the Cum Laude Society (dedicated to honoring scholastic achievement in secondary schools) in 2006, and holds a number of patents related to his research.

Joe Mascetta, M.S., C.A.S.

Joe Mascetta, M.S., C.A.S., taught high school chemistry for 20 years. He also served as the science department coordinator and principal of Mt. Lebanon High School in Pittsburgh, Pennsylvania, a science consultant to the area schools, and was a past president of the Western Pennsylvania Association of Supervision and Curriculum Development (ASCD) and the State Advisory Committee of ASCD.

CONTENTS

How to Use This Book

Barron's Chemistry Practice Plus is designed to offer essential review of key topics and loads of online practice to help you excel in chemistry.

Online Practice

Access more than 400 questions in online quizzes arranged by topic for customized practice! All questions include answer explanations.

What Will You Learn in the Book?

Key review and topics are covered so you can study the essentials needed to succeed.

Learning objectives are listed at the start of each chapter. This list of key ideas will help guide your learning and study plan and allow you to easily return to topics that you want to review again.

Examples are included to help support your learning of surrounding topics.

Tips are given throughout the book to offer helpful notes, reminders, and strategies to improve your learning.

CHAPTER 1

Introduction to Chemistry

Learning Objectives

In this chapter, you will learn how to:

- Distinguish types of matter: i.e., elements, mixtures, compounds, and pure substances.
- Identify chemical and physical properties and changes.
- Explain how energy is involved in these changes.
- Identify and use the SI units of measurements.
- Do mathematical calculations by using scientific notation, dimensional analysis, and proper significant figures.

Chemistry is the study of matter and how it changes. Changes in matter are always accompanied by changes in the energy associated with the matter that is changing. Therefore, chemistry investigates matter and energy as well as the changes they undergo as processes unfold. A good understanding of both matter and energy is thus required to understand the world from a chemical perspective.

Matter

Matter is defined as anything that occupies space and has mass. **Mass** is the quantity of matter that a substance possesses and, depending on the gravitational force acting on it, has a unit of weight assigned to it. Its formula is $w = mg$, where m is the mass of the substance and g is a gravitational constant. Although **weight** can vary as the gravitational constant does, the mass of the body is a constant and can be measured by its resistance to a change of position or motion. This property of mass to resist a change of position or motion is called **inertia**. Since matter does occupy space, we can compare the masses of various substances that occupy a particular unit volume. This relationship of mass to a unit volume is called the **density** of the substance. It can be shown in a mathematical formula as $D = \frac{m}{V}$. The unit of mass (m) commonly used in chemistry is the gram (g), and of volume (V) is the cubic centimeter (cm^3), milliliter (mL), or liter (L).

An example of how density varies can be shown by the difference in the volumes occupied by 1 gram of a metal, such as gold, and 1 gram of Styrofoam. Both have the same mass, 1 gram, but the volume occupied by the Styrofoam is much larger. Therefore, the density of the gold is much larger than that of the Styrofoam. In chemistry, the typical unit for the density of

TIP
Matter occupies space and has mass.

TIP
Density = $\frac{\text{Mass}}{\text{Volume}}$

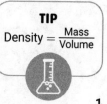

solids is grams/cubic centimeter and for that of liquids is grams/milliliter. Both of these are temperature dependent. The typical unit for the density of gases is grams/liter at standard temperature and pressure. However, the density of gases is usually much lower than the densities of solids and liquids, which is the reason for the unit of liter in the denominator. This ensures a reasonable amount of mass will be reported for gases. Density is an important differentiating property of matter. It is a quantitative value that can be assigned to all matter based on the definition of matter: anything that has mass and volume.

States of Matter

Matter occurs in three states: solid, liquid, and gas. A **solid** has both a definite size and a definite shape. A **liquid** has a definite volume but takes the shape of the container, and a **gas** has neither a definite shape nor a definite volume. These states of matter can often be changed by the addition or subtraction of heat energy. An example is ice changing to liquid water and finally steam.

Chemical and Physical Properties

Physical properties of matter are those properties that can usually be observed with our senses. They include everything about a substance that can be noted when no change is occurring in the type of structure that makes up its smallest component. Some common examples are physical state, color, odor, solubility in water, density, melting point, mass, and weight. All of the properties just listed, except for mass and weight, are independent of the size of the sample of matter being described and are referred to as **intensive**. The intensive nature of a property does not preclude it from being quantitative, however. Solubility, density, and melting point are examples of intensive properties. They are measured properties and contain numbers in their description. Mass and weight are also quantitative properties of matter but depend on the size of the sample. These types of properties are referred to as **extensive**.

Chemical properties are those properties that can be observed in regard to whether or not a substance changes chemically, often as a result of reacting with other substances. Some common examples are: iron rusts in moist air, nitrogen does not burn, gold does not rust, sodium reacts with water, silver does not react with water, and water can be decomposed by an electric current.

Classification of Matter

Chemical and physical properties can be used to classify matter. A common scheme first considers the uniformity of the properties in a sample of matter. If the properties of the sample are distributed evenly, the substance is said to be **homogeneous**. In contrast, substances described as **heterogeneous** reveal an uneven distribution of properties. Examples of homogeneous matter include iron, sulfur, carbon dioxide, or water. In the case of iron, the sample is uniformly gray, is magnetic, and exhibits a density of 7.9 g/cm^3. In the case of sulfur, the sample is uniformly yellow, is not magnetic, and exhibits a density of 2.0 g/cm^3. One example of heterogeneous matter is iron with sulfur on top of it in a test tube. Another example is a bowl containing milk with cereal bits floating on its surface. In both of these cases, the properties of the sample are different in various locations in the sample. In the sulfur/iron example, the bottom is gray, magnetic, and of higher density than the top, which is yellow, not magnetic, and of lower density. This sample represents a physical blend of two materials whose overall composition could be varied by adding or taking away some of either part in this combination. Additionally, the properties of those parts are retained in the various sections of the sample. Physical blends of materials, in no particular ratio, in which the properties of the combining substances are maintained are called **mixtures**. The nonuniform mixture of iron and sulfur in the test tube could be made more uniform and even become homogeneous by vigorously shaking the test tube. Consequently, mixtures could also be homogeneous samples of matter. A familiar type of homogeneous mixture is a solution. As a mixture, solutions have no set constituent proportions but, instead, often display a limited range of composition.

Not all homogeneous matter is a mixture, however. Another type of substance has a uniformity concerning its properties. In this type of matter, the proportions of the substance(s) that make it up are fixed and cannot vary. These are **pure substances**. Iron and sulfur are each examples of one kind of pure substance, while carbon dioxide and water are each examples of another kind of pure substance. Substances like iron or sulfur are pure because they are 100% of one thing and are called **elements**. Substances like carbon dioxide and water are also pure because they contain parts that are put together in particular ratios by mass. This type of pure substance is called a **compound**. Compounds follow one of the basic laws of chemistry called the **Law of Definite Composition** proposed by Joseph Proust in the late eighteenth century. Like mixtures, compounds are blends of materials. However, compounds are specific chemical combinations. This type of interaction requires the merger of specific quantities of the elements that make up the compound. The elements iron and sulfur can be combined chemically (by heating) to make the compound iron(II) sulfide but only in the mass ratio of 63.5% iron and 36.5% sulfur. Additionally, the set of properties associated with a compound is different from the properties of the elements that make it up. Although the compound iron(II) sulfide contains iron, it is not magnetic.

Mixtures should be recognized as physical blends of elements and/or compounds. As such, they could be separated rather easily by taking advantage of the retained properties of the mixture parts. The iron/sulfur mixture previously referenced could easily be separated by passing it by a magnet. The iron would cling to the magnet, while the sulfur would not. A solution (recall it as being a homogeneous mixture) made of water and alcohol is often separated based on the boiling point differences of the water and the alcohol. This is a process called distillation. Filtration is a physical separation technique that takes advantage of solubility differences, and chromatography is a physical separation technique that utilizes chemical affinity differences.

Compounds should be recognized as chemical blends of elements. As such, they could still be separated but would require a chemical process to do so. The compound water referred to previously is a chemical blend of hydrogen and oxygen. This combination could be undone by passing an electric current through a sample of water. The experimental setup for the electrolysis of water is shown in Figure 1.1.

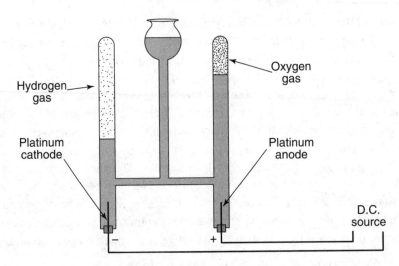

FIGURE 1.1 The Electrolysis of Water

Notice that the hydrogen produced on the left is twice the volume of the oxygen produced on the right. This is reflective of the formula for water, H_2O. It is also important to recognize that the properties of the hydrogen and oxygen products are different than the properties of the water itself. Noteworthy is the difference in phase between the compound water and the elements hydrogen and oxygen that make up the water.

Chemical and Physical Changes

The separation techniques previously described for mixtures and compounds more broadly involve physical and chemical changes that can happen to matter. In general, a **physical change** alters some aspect of the physical properties of matter, but the composition remains constant. The most often altered properties are form and state. Some examples of physical changes are breaking glass, cutting wood, melting ice, and magnetizing a piece of metal. In some cases, the process that caused the change can be easily reversed and the substance regains its original form. Water changing its state is a good example of a physical change. In the solid state, ice, water has a definite size and shape. As heat is added, it changes to the liquid state, where it has a definite volume but takes the shape of the container. When water is heated above its boiling point, it changes to steam. Steam, a gas, has neither a definite size, because it fills the containing space, nor shape, because it takes the shape of the container.

Chemical changes are changes in the composition and structure of a substance. In other words, when a chemical change occurs, new substances are formed. Some examples of chemical change are the rusting of iron, the burning of wood, or the reacting of an acid with a base. These changes are generally not as easily reversed as are physical changes.

Conservation of Mass

When ordinary chemical changes occur, the mass of the reactants equals the mass of the products. This can be stated another way: In a chemical change, matter can neither be created nor destroyed, but only changed from one form to another. This is referred to as the **Law of Conservation of Matter** and was proposed by Antoine Lavoisier in the late eighteenth century (similar in time to the proposal of the Law of Definite Composition previously described). In large part, because of the specificity associated with each of these laws, and more

TIP
A physical change does not alter the identity of the substance. Chemical change does.

particularly the fixed nature of matter that each describes, a recognition of the discreteness of matter emerged. In the very early nineteenth century, the so-called Father of Chemistry, John Dalton, postulated the particulate nature matter displayed by these laws. The concept of an **atom** became accepted on experimental grounds. An atom is the basic unit of an element. The word *atom* is derived from the Greek word *atomos*, which means "indivisible" in that language.

Unlike mixtures and compounds, elements are those substances that cannot be broken down by any means (physical or chemical). The reason for this is made clear by describing elements as composed of only one kind of atom. Compounds could then be viewed as chemical combinations containing more than one kind of atom. This explains the chemical formulas associated with compounds. For example, CO_2 is the formula for the compound carbon dioxide and H_2O is the formula for the compound water. The smallest particle of carbon dioxide is a grouping of one carbon atom and two oxygen atoms. That of water is a grouping of two hydrogen atoms and one oxygen atom. In both examples, there are three atoms with two different kinds of atoms in the grouping. Groupings of atoms like this are referred to as **molecules**. When the grouping contains more than one kind of atom, the molecule is associated with a compound, like CO_2 or H_2O. When the grouping contains only one kind of atom, the molecule is said to be elemental, like H_2 (elemental hydrogen) or O_2 (elemental oxygen). The vast majority of elements are thought of as existing monatomically. For example, the elements carbon and iron are generally described simply by their chemical symbols C and Fe, respectively.

A summary of the classification of matter previously described is seen in Figure 1.2.

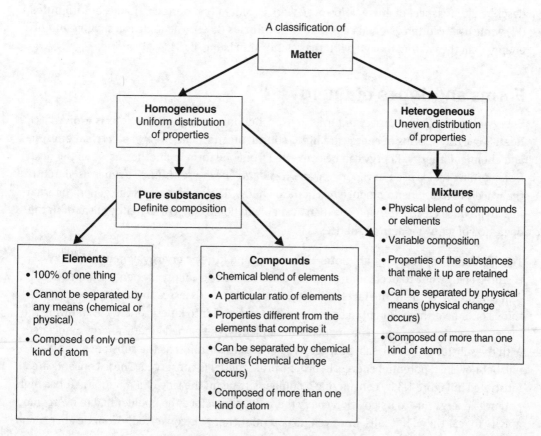

FIGURE 1.2 A Classification of Matter

Energy

The concept of energy plays an important role in all of the sciences. In chemistry, all physical and chemical changes have energy considerations associated with them. To understand how and why these changes happen, an understanding of energy is required.

Definition of Energy

Energy is defined as the capacity to do work or heat things. Work is done whenever a force is applied over a distance. Therefore, anything that can force matter to move, to change speed, or to change direction has energy. Heat is the flow of energy due to temperature differences. The following example will help you understand this definition of energy.

Raising a hammer above a nail about to be pounded into a block of wood stores energy in the hammer. Energy is also stored in the muscles of the arm that has lifted the hammer. When the hammer is brought down to impact upon the nail, the nail will likely be forced to move into the wood and so work is done on the nail. Continuing to pound on the nail will force it to move even farther into the wood but may also result in noticeable heating of the nail. If the nail is unable to move farther into the wood upon additional pounding (perhaps because the wood is very hard), the only thing that will occur is that the nail will heat up. In other words, upon touching the nail with your cooler fingers, the nail will be able to transfer energy to you with a greater capacity than it previously had before being pounded.

Typical units for energy include the **joule** (J) and the **calorie** (cal). Each could be used to describe either aspect of energy. However, work is generally reported in joules or **kilojoules** (kJ) while heat is often given in calories or **kilocalories** (kcal). The joule is a smaller unit of energy than the calorie, though, with the relationship being 4.184 J = 1 calorie.

Forms and Types of Energy

Energy may appear in a variety of forms. Most commonly, energy in reactions is evolved as **heat**. Some other forms of energy are **light, sound, mechanical energy, electrical energy**, and **chemical energy**. Energy can be converted from one form to another, as when the heat from burning fuel is used to vaporize water to steam. The energy of the steam is used to turn the wheels of a turbine to produce mechanical energy. The turbine turns the generator armature to produce electricity, which is then available in homes for use as light or heat or in the operation of many modern appliances.

Two general types of energy are **potential energy** and **kinetic energy**. Potential energy is stored energy due to overcoming forces in nature. Kinetic energy is energy of motion. In other words, by virtue of the fact that matter is moving, it possesses the ability to move other things or to heat them. Potential energy can be converted into kinetic energy by allowing the forces that were overcome to be applied. Consequently, the gained kinetic energy associated with the substance has the same capacity to move or heat things as the substance's original, but now lost, potential energy. In other words, energy is conserved. These concepts are illustrated in Figure 1.3 by considering a boulder sitting on the side of a mountain. It has high potential energy due to its position above the valley floor since the boulder had to overcome gravity to rest there. If it falls, however, its potential energy is converted to kinetic energy and will move or heat up the matter with which it collides.

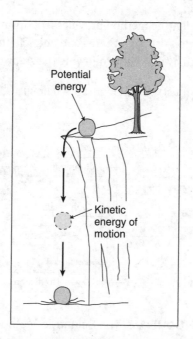

FIGURE 1.3 Conversion of Potential Energy to Kinetic Energy

Conservation of Energy

Experiments have shown that energy is neither gained nor lost in physical or chemical changes. This principle is known as the **Law of Conservation of Energy** and is often stated as follows: Energy is neither created nor destroyed in ordinary physical and chemical changes. If the system under study loses energy, the reaction is exothermic and the ΔH is negative. Therefore, the system's surroundings must gain the energy that the system loses so that energy is conserved.

Conservation of Mass and Energy

With the introduction of atomic theory and a more complete understanding of the nature of both mass and energy, it was found that a relationship exists between these two concepts. Einstein formulated the **Law of Conservation of Mass and Energy**. This states that mass and energy are interchangeable under special conditions. The conditions have been created in nuclear reactors and accelerators, and the law has been verified. This relationship can be expressed by Einstein's famous equation:

$$E = mc^2$$

$$\text{Energy} = \text{Mass} \times (\text{Velocity of light})^2$$

Scientific Method

In order to classify matter, chemists must observe the properties of the matter they investigate. This act of observation is typically the first step scientists make in discovering the aspects of their discipline. As such, observation is generally recognized as the starting point in the **scientific method**. The scientific method provides a pathway by which scientists investigate and learn about the world. Although there is no one scientific method, most scientific understanding has been gained by generally following the steps outlined here.

The stages of this process are illustrated in Figure 1.4 below:

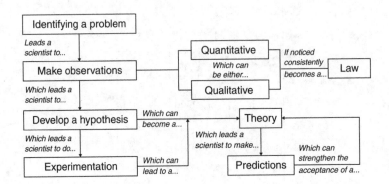

FIGURE 1.4 The Scientific Method

Measurements and Calculations

As previously discussed, a vital part of scientific endeavor involves making quantitative observations or measurements. Every measurement has two aspects: a number and a unit. Once measurements are made, they are typically used to express, explain, or exemplify scientific ideas. In addition, once made they are often mathematically manipulated to better describe the concept under investigation. Therefore, knowing the measurement system typically used, applying a straightforward approach in handling the numbers (which could be very large or very small), and understanding mathematical machinations that may be required are all critical for success.

The International System of Units (SI)

It is important that scientists around the world use the same units when communicating information. For this reason, scientists use the modernized metric system, designated in 1960 by the General Conference on Weights and Measures as the International System of Units. This is commonly known as **SI**, an abbreviation for the French name Le Système International d'Unités. It is now the most common system of measurement in the world. There are minor differences between the SI and metric systems. For the most part, the quantities are interchangeable.

The reason SI is so widely accepted is due to its straightforward, simple approach. The system includes only seven base units that can be inflated or deflated by the use of prefixes related by a decimal. Quantities other than those described by the base units can be derived by mathematical manipulation of the base units or other derived units.

The six base units that can be used to express the fundamental quantities of measurement are specifically defined within the system.

TABLE 1.1 SI Base Units

Quantity	Unit	Abbreviation
mass	kilogram	kg
length	meter	m
time	second	s
electric current	ampere	A
temperature	kelvin	K
amount of substance	mole	mol

If the magnitude of an SI unit is not appropriate for the object being measured, prefixes can be combined with the unit to adjust the unit's size. The prefixes represent multiples or fractions of 10. Table 1.2 gives some prefixes commonly used in the SI system.

TABLE 1.2 Prefixes Used with SI Units

Prefix	Symbol		Meaning	Exponential Notation
exa-	E		1,000,000,000,000,000,000	10^{18}
peta-	P		1,000,000,000,000,000	10^{15}
tera-	T		1,000,000,000,000	10^{12}
giga-	G		1,000,000,000	10^{9}
mega-	M		1,000,000	10^{6}
kilo-	k		1,000	10^{3}
hecto-	h		100	10^{2}
deka-	da		10	10^{1}
—	—		1	10^{0}
deci-	d	Commonly Used Prefixes	0.1	10^{-1}
centi-	c		0.01	10^{-2}
milli-	m		0.001	10^{-3}
micro-	μ		0.000 001	10^{-6}
nano-	n		0.000 000 001	10^{-9}
pico-	p		0.000 000 000 001	10^{-12}
femto-	f		0.000 000 000 000 001	10^{-15}
atto-	a		0.000 000 000 000 000 001	10^{-18}

Because of the prefix system, units can be easily related by some factor of 10. Below is a list of some SI unit equivalents for various fundamental quantities.

Length
10 millimeters (mm) = 1 centimeter (cm)
100 centimeters (cm) = 1 meter (m)
1,000 meters (m) = 1 kilometer (km)

Time
1,000,000 microseconds (μs) = 1 second (s)
10 deciseconds (ds) = 1 second (s)
1,000,000,000 nanoseconds (ns) = 1 second (s)

Mass
1,000 milligrams (mg) = 1 gram (g)
1,000 grams (g) = 1 kilogram (kg)

There are some interesting relationships between volume and mass units in the SI system. Because water is most dense at 4°C, the gram was intended to be 1 cubic centimeter of water at this temperature.

Temperature Measurements

The most commonly used temperature scale in scientific work is the **Celsius** scale. For a long time it was called the centigrade scale because it is based on the concept of dividing the distance on a thermometer between the freezing point of water and its boiling point into 100 equal markings or degrees.

The SI temperature scale and unit are related to the Celsius scale and unit. The SI temperature scale is based on the lowest theoretical temperature (called absolute zero). This temperature has never actually been reached, but scientists in laboratories have reached temperatures within about a 100 trillionths of a degree above absolute zero. Sir William Thomson, also known as Lord Kelvin, proposed this scale on which a unit is the same size as a Celsius degree but where the zero mark has been displaced lower. Consequently, it is referred to as the **Kelvin** temperature scale. In the Kelvin scale, the magnitude of the temperature value is directly proportional to the average kinetic energy of the sample. That's why the Kelvin scale is often called the absolute temperature scale. The direct relationship between temperature and energy is not the case when using the Celsius scale, in which negative numbers and zero are commonly measured values. For that reason, it is common in calculations involving temperature that values measured in Celsius be converted to the Kelvin scale.

Through experiments and calculations, it has been determined that absolute zero is 273.15 degrees below zero on the Celsius scale. This figure is usually rounded off to −273°C.

Figure 1.5 displays the graphic and algebraic relationships among three temperature scales: the Celsius and Kelvin, commonly used in chemistry, and the Fahrenheit.

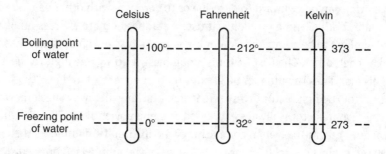

Conversion Formulas

$$K = °C + 273$$
$$°C = K - 273°$$

FIGURE 1.5 Temperature Scale Relationships

Derived Quantities and Units

Since there are only seven SI base units but many more than seven quantities that need to be described by a unit, most SI units are constructed through mathematical manipulation of related units. These units are said to be derived. A derived unit is comprised of base or other derived units put together in some mathematical operation that describes the considered quantity. Often, the combination of units is given a name to simplify the expression of the unit. It is common for the name to have been given in honor of a famous scientist who worked in the particular branch of science for which the derived unit is commonly used. Table 1.3 contains some commonly derived quantities, how their units are derived, as well as (perhaps) the name given to the unit.

TABLE 1.3 Derived Quantities and Units

Quantity	Derivation by Quantity	Derivation(s) by Unit	Name
area	length $\times$ length	m^2, cm^2, or dm^2	—
volume	length $\times$ length $\times$ length	m^3, cm^3, or dm^3	liter (L) = dm^3
density	mass/volume	g/cm^3, g/mL, or g/L	—
velocity	distance/time	m/s	—
acceleration	distance/time^2	m/s^2	—
force	mass $\times$ acceleration	$kg \times m/s^2$	newton (N)
energy	force $\times$ distance	$N \times m$ or $kg \times m^2/s^2$	joule (J)

Scientific Notation

Often the magnitude of a measurement is very large or very small. So the numbers used to express the value can be awkard to describe or to compare with other values. Scientific notation alleviates both of these problems. It uses a standard way to express numbers that easily allows the numbers to be compared. A number written in scientific notation has two factors multiplied by each other. The first factor is a number with only one digit to the left of the decimal. This mandates the number be greater than or equal to 1 and less than 10. In this way, the magnitude of the first factor in any scientific notation will always be somewhat similar in value. The second factor is always expressed as some power of 10. Since the first factor is always similar in magnitude, when comparing scientific notation numbers the exponent on the base 10 can easily be used to compare the sizes of the numbers expressed.

Scientific Notation

First factor × Second factor

(Some number with 1 digit to left of decimal) × (some power of 10)

EXAMPLES

3,400,000 in scientific notation is 3.4×10^6

0.0000034 in scientific notation is 3.4×10^{-6}

With large numbers, such as 3,400,000, first move the decimal point to the left until only one digit remains to the left of it (3.400000). Then indicate the number of moves of the decimal point as the exponent of 10 (3.4×10^6). With a very small number such as 0.0000034, first move the decimal point to the right until only one nonzero digit is to the left of it (0000003.4). Then indicate the number of moves as the negative exponent of 10 (3.4×10^{-6}).

Precision, Accuracy, and Uncertainty

Two other factors to consider in measurement are **precision** and **accuracy**. Precision indicates the reliability or reproducibility of a measurement. Accuracy indicates how close a measurement is to its known or accepted value.

For example, suppose you were taking a reading of the boiling point of pure water at sea level on a normal day. Using the same thermometer in three trials, you record 96.8, 96.9, and 97.0 degrees Celsius. Since these figures show a high reproducibility, you can say that they are precise. However, the values are considerably off from the accepted value of 100 degrees Celsius, so you say they are not accurate. In this example, you would probably suspect that the inaccuracy was the fault of the calibration of the thermometer.

Regardless of precision and accuracy, all measurements have a degree of **uncertainty**. This is usually dependent on one or both of two factors—the limitation of the measuring instrument and the skill of the person making the measurement. Uncertainty can best be shown by example.

Significant Figures

Any time a measurement is recorded, it includes all the digits that are certain plus one uncertain digit. These certain digits plus the one uncertain digit are referred to as **significant**

figures. These digits are all *reasonably* certain because you absolutely know all but the last and have made a guess based on reason for that last one. The more digits you are reasonably able to record in a measurement, the less relative uncertainty there is in the measurement. If a measurement is made and a nonzero value is reported in a place, it had to have been reasonably measured or a value would not be there. The value of 0 may be interpreted in different ways, however. Sometimes it is a significant figure (meaning it is in a measured place), but other times it is not a significant figure (meaning it is simply a placeholder giving proper magnitude to a number). Table 1.4 summarizes the rules of significant figures.

TABLE 1.4 Zero Rules for Significant Figures

Rule	Example	Number of Significant Figures
All digits other than zeros are significant.	25 g	2
	5.471 g	4
Zeros between nonzero digits are significant. (They had to have been in absolutely measured places.)	309 g	3
	40.06 g	4
Final zeros to the right of the decimal point are significant. (The don't need to be there to express the magnitude of the number; therefore, they must have been measured.)	6.00 mL	3
	2.350 mL	4
In numbers smaller than 1, zeros to the left or directly to the right of the decimal point are not significant.	0.05 cm	1 — The zeros merely mark the position of the decimal point.
	0.060 cm	2 — The first two zeros mark the position of the decimal point. The final zero is significant.

One last rule deals with final zeros in a whole number. These zeros may or may not be significant, depending on the measuring instrument. For instance, if an instrument that measures to the nearest mile is used, the number 3,000 miles needs to have four significant figures. If, however, the instrument in question records miles to the nearest thousands, there should be only one significant figure. The number of significant figures in 3,000 could be one, two, three, or four, depending on the limitation of the measuring device.

This problem can be avoided by using scientific notation. By convention, the first factor in any scientific notation number used to express a measurement contains only significant figures. For this example, the following notations all indicate different numbers of significant figures:

3×10^3 one significant figure

3.0×10^3 two significant figures

3.00×10^3 three significant figures

3.000×10^3 four significant figures

Another common way to make zeros significant, when they would not normally be considered so, is to place a decimal point at the end of the whole number when it would not normally be shown. For example:

3000   (one significant figure)

3000.   (four significant figures)

Calculations with Significant Figures

When you do calculations involving numbers that do not have the same number of significant figures in each, keep the following two rules in mind.

First, in multiplication and division, the number of significant figures in a product or a quotient of measured quantities is the same as the number of significant figures in the quantity having a smaller number of significant figures.

EXAMPLE

Problem	Unrounded answer	Answer rounded to the correct number of significant figures
4.29 cm × 3.24 cm =	13.8996 cm^2 =	13.9 cm^2

Explanation: Both measured quantities have three significant figures. Therefore, the answer should be rounded to three significant figures.

Second, when adding or subtracting measured quantities, the sum or difference should be rounded to the same number of decimal places as the quantity having the fewest decimal places.

EXAMPLE

Problem	Unrounded answer	Answer rounded to the correct number of significant figures
3.56 cm		
2.6 cm		
+6.12 cm		
Total =	12.28 cm =	12.3 cm

Explanation: One of the quantities added has only one decimal place. Therefore, the answer should be rounded to only one decimal place.

Dimensional Analysis

Dimensional analysis is a mathematical technique often used by scientists to convert a quantity given in one unit to the same quantity expressed in a different unit. This may eliminate the need to express the number in scientific notation if it is too big or too small to work with easily, thereby expressing the measurement in an appropriately sized unit. For example, you may alter the measured mass of a large sample of iron given in milligrams to kilograms so that it may be more easily compared to other large samples of metals for which you know their mass in kilograms.

To use dimensional analysis, an equation needs to be written that mathematically changes the units of the given value to that of the unknown value. The first part of the dimensional analysis equation begins with the given quantity itself. Next, a fraction that equals 1 is multiplied by the given quantity. In the fraction, the desired units are in the numerator and the units of the given quantity are in the denominator. Since the fraction is equal to 1, the given quantity itself stays the same. The example that follows shows how 155,500 milligrams can be expressed in kilograms. This may be a more appropriate expression of the given quantity based on the quantity's size.

$$155{,}500 \text{ mg} \times \frac{1 \text{ kg}}{1{,}000{,}000 \text{ mg}} = 0.1555 \text{ kg}$$

Analysis of the equation shows that the mg unit in the given quantity is canceled out by the mg unit in the denominator of the fraction. The unit that does not cancel is then associated with the answer. The mathematical operation is then applied to the numbers to produce the value in front of that unit.

CHAPTER 2

Atomic Structure and the Periodic Table of the Elements

Learning Objectives

In this chapter, you will learn how to:

- Describe the history of the development of atomic theory.
- Explain the structure of atoms, their main energy levels, sublevels, orbital configuration, and the rules that govern how they are filled.
- Place atoms in groups and periods based on their atomic structure.
- Explain how chemical and physical properties are related to positions in the Periodic Table, including atomic size, ionic size, electronegativity, acid-forming properties, and base-forming properties.
- Explain the nature of radioactivity, the types and characteristics of each, and the inherent dangers.
- Identify the changes that occur in a decay series.

The idea of small, invisible particles being the building blocks of matter can be traced back more than 2,000 years to the Greek philosophers Democritus and Leucippus. These particles, considered to be so small and indestructible that they could not be divided into smaller particles, were called **atoms**, which is related to the Greek word for indivisible. This early concept of atoms was not based upon experimental evidence but was simply a result of thinking and reasoning on the part of the philosophers. Not until the late eighteenth century did experimental evidence in favor of the atomic hypothesis begin to accumulate. This evidence helped formulate two fundamental chemical laws: the Law of Conservation of Matter and the Law of Definite Composition. Each of these hinted at the discreteness of matter. This specificity strongly suggested that matter was particulate. Finally, around 1805, John Dalton proposed some basic assumptions about atoms based on those laws and his own Law of Multiple Proportions. These assumptions are very closely related to what scientists presently know about atoms. The basic postulates of **Dalton's atomic theory** are summarized here:

1. All matter is made up of very small, discrete particles called atoms.
2. All atoms of an element are alike in weight, and this weight is different from that of any other kind of atom.
3. Atoms cannot be subdivided, created, or destroyed.

4. Atoms of different elements combine in simple whole-number ratios to form chemical compounds.

5. In chemical reactions, atoms are combined, separated, or rearranged.

The model that emerges from Dalton's atomic theory is that atoms can be viewed as hard, little spheres. According to Dalton, the only difference between an atom of iron and an atom of zinc was its mass. He thought that an atom of zinc was a hard, little sphere that was simply more massive than that of iron. It appeared as if mass was integral in determining the properties of the atoms. Although this model was able to explain many observations, as the nineteenth century concluded scientists began to recognize there were significant phenomena that Dalton's theory could not explain. Particularly, an explanation for the apparent electrical nature of matter was an issue that needed to be addressed. Electricity was used in many experiments to promote chemical change. Many elements were discovered by running electricity through the compounds in which they were bound. About 100 years after Dalton's theory was introduced, it became clear that the atom was somewhat misnamed as experimentation proved that the atom was indeed composed of internal parts and was not indivisible after all.

Early Atomic Models

From around the beginning of the twentieth century, scientists have been gathering evidence about the structure of atoms and fitting the information into a model describing atomic structure.

J. J. Thomson and the Plum Pudding Model

The discovery of the electron as the first subatomic particle is credited to **J. J. Thomson** (England, 1897). He used an evacuated tube connected to a spark coil as shown in Figure 2.1. As the voltage across the tube was increased, a beam became visible. This was referred to as a cathode ray. Thomson found that the beam was deflected by both electrical and magnetic fields. However, since magnetic and electric fields were known not to bend light and the beam bent toward the positive end of the electric field, he concluded that cathode rays are made up of very small, negatively charged particles of matter, which became known as **electrons**.

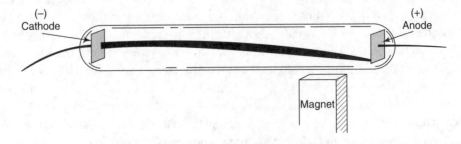

FIGURE 2.1 Cathode Ray Tube

By measuring the deflection angle of the cathode ray in combination with a knowledge of the strength of the fields used, Thomson was able to determine the electrical charge on the electron compared to its mass. This was called the charge-to-mass ratio, e^-/m, and was equal to

1.76×10^8 coulombs per gram. This was a major step toward understanding the nature of the particle. Thomson was awarded a Nobel Prize in 1906 for his accomplishment.

It was an American scientist, **Robert Millikan**, who in 1909 was able to measure solely the charge on an electron.

Millikan was able to calculate the charge of the electron as 1.6×10^{-19} coulombs. By combining this information with the results of Thomson, he could calculate a value for the mass of a single electron. Eventually, this number was found to be 9.11×10^{-28} grams, about 2,000 times lighter than the smallest atom, hydrogen.

Since atoms were known to be neutral particles, Thomson imagined the atom as being composed of the lightweight, negatively charged electrons embedded in a diffuse sphere of positiveness. The balance of charge between the negative electrons and the positive sphere allowed for the atom's overall neutral charge. He likened the atom to an English dessert called plum pudding. In the model, the electrons were analogous to the plums and the diffuse positive sphere similar to the pudding into which the plums (electrons) were embedded.

Ernest Rutherford and the Nuclear Model

In an attempt to verify the plum pudding model, **Ernest Rutherford**, in England in 1911, performed an experiment in which a beam of alpha radiation was directed at a very thin sheet of gold foil (see Figure 2.2). Alpha radiation is a type of nuclear decay, discovered previously by Rutherford, that is positively charged and relatively heavy. Expecting that all of the alpha particles would shoot straight through the gold atom's diffuse, positively charged pudding in which its lightweight electrons were embedded, Rutherford was astounded when about 1 in every 8,000 particles experienced a deflection. Even more interesting, about 1 in every 20,000 particles were deflected almost directly back at the source.

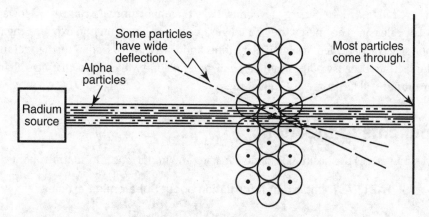

FIGURE 2.2 Rutherford's Experiment

Rutherford reasoned that the rebounded alpha particles must have experienced some powerful force within the atom. He also assumed that the source of this force must occupy a very small amount of space because so few alpha particles had been deflected. Rutherford concluded that the source must be a densely packed bundle of matter with a positive charge. He called this positive bundle the **nucleus**. He further surmised that the volume of a nucleus was very small compared with the total volume of an atom because most of the alpha particles passed through unaffected.

Further experiments showed that the nucleus was made up of still smaller particles called **protons**. Rutherford realized, however, that protons by themselves could not account for the entire mass of the nucleus. He predicted the existence of a new nuclear particle that would be neutral and would account for the missing mass. In 1932, James Chadwick discovered this particle, the **neutron**.

Niels Bohr and the Planetary Model

Although the Rutherford model for the atom was able to explain the scattering of alpha particles in the gold foil experiment, it was unable to explain other behavior associated with matter. The most noteworthy is that when supplied energy, either in the form of heat or electricity, matter releases specific types of light. When copper is placed into a flame, the flame displays a green color to the naked eye. Sodium produces a yellow flame, and lithium produces a red flame. A glass bulb filled with hydrogen, through which electricity is run, glows with a pink color. One filled with neon glows orange. If the light emitted is analyzed with a prism to reveal the components of the light, it is shown to be a series of very specific lines of color. The series of lines realized by this resolution is referred to as a bright-line spectrum. It was known that each element exhibited a unique bright-line spectrum that can be used to identify the element. Therefore, a new model for the atom was needed to explain this behavior.

In 1913, **Niels Bohr** (Denmark) proposed his model of the atom, which was capable of explaining bright-line spectra. He pictured the atom as having a dense, positively charged nucleus and negatively charged electrons in specific spherical orbits, also called energy levels or shells, around this nucleus. These energy levels are arranged concentrically around the nucleus, and each level is designated by a number: 1, 2, 3, . . . The closer to the nucleus, the less energy an electron needs in one of these levels. However, an electron has to gain energy to go from one level to another that is farther away from the nucleus.

Electrons excited to a higher energy level may have the opportunity to return to a lower energy level by releasing the energy they no longer need. This phenomenon explains the unique bright-line spectra identified for particular atoms and will be discussed in further detail in an upcoming section. Because of its simplicity and general ability to explain chemical change, the **Bohr model** still has some usefulness today.

Components of the Atom

Table 2.1 summarizes the basic subatomic particles and important information about them.

TABLE 2.1 Important Information about Subatomic Particles

Particle	Charge	Symbol	Actual Mass	Relative Mass Compared to Proton	Discovery
Electron	$- (e^-)$	$_{-1}^{0}e^-$	9.109×10^{-28} g	1/1,837	J. J. Thomson–1897
Proton	$+ (p^+)$	$_{1}^{1}P^+$	1.673×10^{-24} g	1	E. Rutherford–1917
Neutron	$0 (n^0)$	$_{0}^{1}n^0$	1.675×10^{-24} g	1	J. C. Chadwick–1932

When these components are used in the model, the protons and neutrons are shown in the nucleus. These particles are known as **nucleons**. The electrons are shown outside the nucleus.

The number of protons in the nucleus of an atom determines the **atomic number**, symbolized Z. All atoms of the same element have the same number of protons and therefore the same atomic number; atoms of different elements have different atomic numbers. Thus, the atomic number identifies the element.

The sum of the number of protons and the number of neutrons in the nucleus is called the **mass number**, symbolized by the letter A. A particular atom with a specific number of protons and neutrons is called a **nuclide**.

As a result of having a different number of neutrons, different nuclides of the same element, called **isotopes**, have different masses.

The isotopes of a particular element all have the same number of protons and electrons but different numbers of neutrons. Most elements are found naturally as a mixtures of isotopes, some stable and others radioactive. Tin, for example, has ten stable isotopes, the most of any element.

The percentage of each isotope for a naturally occurring element on Earth is nearly always the same, no matter where the element is found. The percentage at which each of an element's isotopes occurs in nature is taken into account when calculating the element's **atomic mass**. The atomic mass for an element found on the Periodic Table is the weighted average of the atomic masses of the naturally occurring isotopes of an element.

> **TIP**
> Isotopes have the same atomic number but different mass numbers. This means they differ in the number of neutrons, not protons.

Calculating Average Atomic Mass

The average atomic mass of an element depends on both the mass and the relative abundance of each of the element's isotopes. For example, naturally occurring copper consists of 69.17% copper-63, which has an atomic mass of 62.919598 amu, and 30.83% copper-65, which has an atomic mass of 64.927793 amu. The average atomic mass of copper can be calculated by multiplying the atomic mass of each isotope by its relative abundance (expressed in decimal form) and adding the results:

$$(0.6917 \times 62.919598 \text{ amu}) + (0.3083 \times 64.927793 \text{ amu}) = 63.55 \text{ amu}$$

Therefore, the calculated average atomic mass of naturally occurring copper is 63.55 amu. Average atomic masses of the elements are, more often than not, listed under the symbol for the element in a Periodic Table as a decimal number. These atomic masses are typically rounded to one decimal place when doing chemical calculations. The whole number, usually above the elemental symbol, is the atomic number, Z. Mass numbers, A, are not shown as information on the Periodic Table because the symbol represents an average atom in the set of isotopes for that element and not any particular isotope.

Mass Spectrometry

The atomic masses for the isotopes of copper listed in the previous section are actually *relative* atomic masses. The mass of an individual atom is too small to determine directly and absolutely. The modern standard to which the mass of every atom is compared is carbon-12. The carbon-12 atom has been assigned a value of 12.0000 amu (atomic mass units). A device, called a mass spectrometer, can measure relative atomic masses compared to carbon-12, as shown in Figure 2.3.

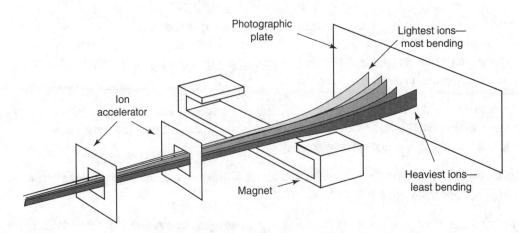

FIGURE 2.3 Mass Spectrometer

Particles of matter to be analyzed are placed into the device and ionized. Ionization knocks electrons out of atoms but doesn't change the mass of the atoms appreciably as each electron has such a low relative mass. The ions are then accelerated past magnetic and electric fields to which they respond. Less massive particles are deflected to a higher degree than more massive particles. The particles are then differentiated by where they land on the detector plate. A display of the differences can be generated and is called a spectrum. The measured differences, as to where each particle lands, allows for relative atomic masses to be determined.

Atomic Spectra

The Bohr model was based on a simple postulate; the electrons around the nucleus in the atom had only certain levels of energy they could possess. Bohr applied this model successfully to the hydrogen atom, which contains a single electron. When the electron is in the lowest energy level, it is said to be in the **ground state** and is closest to the nucleus. When the electron is elevated to any of the higher energy levels (i.e., 1, 2, 3, . . . , ∞), it is said to be excited and is farther away from the nucleus. In going from the ground state to an excited state, the electron must absorb a specific amount of energy. This packet of energy is the difference between the levels of energy between which the electron moves. A bundle of energy like this was referred to by the German scientist Max Planck as a **quantum**. Once in an excited state, the electron can drop down to lower energy levels, this time *releasing* a quantum (i.e., the energy difference between the levels). The formula for a change in energy (ΔE) is:

$$\Delta E_{electron} = E_{final} - E_{initial}$$

Although this bundle of energy can be referred to as a quantum, it is also often called a photon. A photon is a term for a particle of light that was coined by Einstein. The amount of energy in a photon is related to the particle's wavelike nature. That energy can be calculated using the Planck equation, $E = h \times v$, where E is the quantity of energy, h is Planck's constant, and v is the frequency of the light. Frequency can be converted to the wavelength by the speed of light equation, $\lambda \times v = c$, where λ is the wavelength and c is the speed of light in a vacuum.

Important Equations

(Concerning the Particle and Wavelike Nature of Light)

$E = h \times v$ where $h = 6.626 \times 10^{-34}$ J · s (Planck's constant)

$\lambda \times v = c$ where $c = 2.998 \times 10^8$ m/s (speed of light)

Bohr was able to assign values to the various energy levels in the hydrogen atom via his equation:

$$E_n = -1{,}310 \text{ kJ}/n^2$$

In this equation, E_n represents the energy possessed by the electron at the particular energy level designated by the n-value $(1, 2, 3, \ldots, \infty)$. Figure 2.4 shows a chart depicting many of the possible transitions from higher to lower energy levels, for the single electron in hydrogen, based on the energies calculated in the Bohr equation. A few of them have been converted into the wavelength of the radiation released using the equations above.

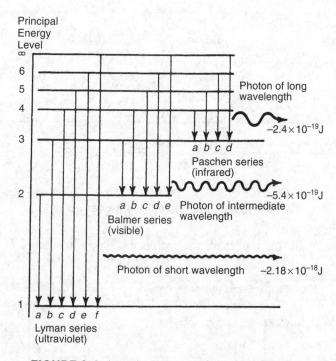

FIGURE 2.4 Atomic Spectra Chart for Hydrogen

The energy values shown were calculated from Bohr's equation.

Each of the first three energy levels has a particular name associated with the emissions that occur when an electron reaches that energy state from levels above it. When an electron drops from a level higher than the first down to $n = 1$, the emissions, consisting of ultraviolet radiation, are known as the **Lyman series**. Note in Figure 2.4 that when an electron drops to the next two higher levels, those emissions have the names **Balmer series** (for $n = 2$) and **Paschen series** ($n = 3$).

Spectroscopy

When the light emitted by energized atoms is examined with an instrument called a **spectroscope**, the prism or diffraction grating in the spectroscope disperses the light. This allows an examination of the emission **spectra**, which is a display of distinct colored lines. Since only particular energy jumps are available in each type of atom, each element has its own unique bright-line spectrum made up of only the lines of a specific wavelength that correspond to its electron structure. The four smallest energy transitions associated with the Balmer series in hydrogen (lines *a* through *d* in Figure 2.4) represent the visible wavelengths of light that can be released by a relaxing electron in that atom. These are the four lines of color in hydrogen's bright-line spectrum. Line *a* is red, line *b* is teal, and lines *c* and *d* are violet in color.

For comparison purposes, a generic visible light spectrum is shown below in Figure 2.5. This diagram displays the colors associated with the various wavelengths of visible light.

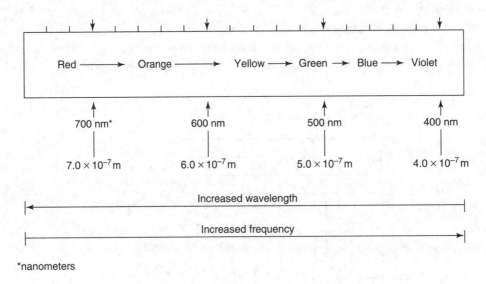

*nanometers

FIGURE 2.5 The Visible Light Spectrum

A partial emission (bright-line) spectrum for hydrogen is shown in Figure 2.6.

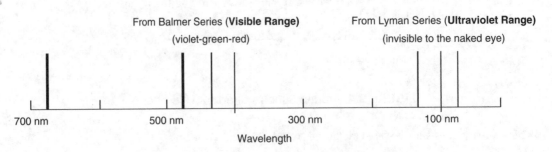

FIGURE 2.6 Partial Emission Spectrum for Hydrogen

The left-hand group is in the visible range and is part of the Balmer series. The right-hand group is in the ultraviolet region and belongs to the Lyman series.

Spectral lines like these can be used in the identification of unknown specimens.

Nuclear Transformations and Stability

At the same time advances in atomic theory were occurring, scientists were noticing phenomena associated with emissions from the nucleus of atoms in the form of X-rays. While Roentgen announced the discovery of X-rays, Becquerel was exploring the **phosphorescence** of some materials. Becquerel's work received little attention until early in 1898, when Marie and Pierre Curie entered the picture.

While searching for the source of the intense radiation in uranium ore, Marie and Pierre Curie used tons of uranium ore to isolate very small quantities of two new elements, radium and polonium, which are both radioactive. Along with Becquerel, the Curies shared the Nobel Prize in Physics in 1903.

The Nature of Radioactive Emissions

While the early separation experiments were in progress, an understanding was slowly being gained about the nature of the spontaneous emissions from various radioactive elements. Becquerel thought at first that they were simply X-rays. However, three different kinds of radioactive emissions, now called **alpha particles**, **beta particles**, and **gamma rays**, were soon found. We now know that alpha particles are positively charged particles of helium nuclei, beta particles are streams of high-speed electrons, and gamma rays are high-energy radiation similar to X-rays. The emissions of these three types of radiation are depicted in Figure 2.7 below.

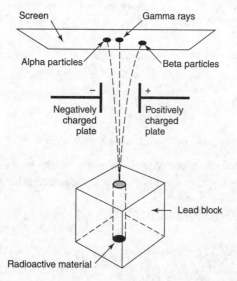

FIGURE 2.7 Deflection of Radioactive Emissions

> **TIP**
> Learn the types of radiation and the characteristics of each.

The important characteristics of each type of radiation can be summarized as follows:

Alpha Particle   (helium nucleus $^4_2$He) Positively charged, 2+

1. Ejection reduces the atomic number by 2 and the mass number by 4.
2. High energy, relatively low velocity.
3. Range: about 5 cm in air.
4. Shielding needed: stopped by the thickness of a sheet of paper, skin.
5. Interactions: produces about 100,000 ionizations per centimeter; repelled by the positively charged nucleus; attracts electrons but does not capture them until its speed is much reduced.
6. An example: thorium-230 has an unstable nucleus and undergoes radioactive decay through alpha emission. The nuclear equation that describes this reaction is:

$$^{230}_{90}\text{Th} \rightarrow {}^4_2\text{He} + {}^{226}_{88}\text{Ra}$$

In a decay reaction like this, the initial element (thorium-230) is called the parent nuclide and the resulting element (radium-226) is called the daughter nuclide.

Beta Particle   (fast electron) Negatively charged, 1−

1. Ejected when a *neutron* decays into a proton and an electron.
2. High velocity, relatively low energy.
3. Range: about 12 m.
4. Shielding needed: stopped by 1 cm of aluminum or the thickness of the average book.
5. Interactions: weak because of high velocity, but produces about 100 ionizations per centimeter.
6. An example: protactinium-234 is a radioactive nuclide that undergoes beta emission. The nuclear equation is:

$$^{234}_{91}\text{Pa} \rightarrow {}^{234}_{92}\text{U} + {}^{\ 0}_{-1}\text{e}$$

Gamma Rays   (electromagnetic radiation identical with light; high energy) No charge

1. Beta particles and gamma rays are usually emitted together; after a beta particle is emitted, a gamma ray follows.
2. Arrangement in nucleus is unknown. Same velocity as visible light.
3. Range: no specific range.
4. Shielding needed: about 13 cm of lead.
5. Interactions: weak of itself; gives energy to electrons, which then perform the ionization.

Methods of Detection of Alpha, Beta, and Gamma Rays

All methods of detection of these types of radiations rely on their ability to ionize. Three methods are in common use.

1. *Photographic plate.* The fogging of a photographic emulsion led to the discovery of radioactivity. If this emulsion is viewed under a high-power microscope, it is seen that beta and gamma rays cause the silver bromide grains to develop in a scattered fashion.

2. *Scintillation counter.* A fluorescent screen (e.g., ZnS) will show the presence of electrons and X-rays, as already mentioned. If the screen is viewed with a magnifying eyepiece, small flashes of light, called scintillations, are observed. By observing the scintillations, one cannot only detect the presence of alpha particles but can also actually count them.

3. *Geiger counter.* This instrument is perhaps the most widely used at the present time for determining individual radiation. Any particle that produces an ion gives rise to an avalanche of ions, so the type of particle cannot be identified. However, each individual particle can be detected.

Decay Series, Transmutations, and Half-Life

The nuclei of radium, uranium, and other radioactive elements are continually disintegrating. The spontaneous disintegration of radium produces the gas known as radon. The time required for half of the atoms of a radioactive nuclide to decay is called its **half-life**.

For example, for radium half of all the radium nuclei present, on average, will have disintegrated to radon in 1,590 years. In another 1,590 years, half of this remainder will decay, and so on. When a radium atom disintegrates, it loses an alpha particle. Upon gaining two electrons, that alpha particle eventually becomes a neutral helium atom. The remainder of the atom becomes radon.

Such a conversion of an element to a new element (because of a change in the number of protons) is called a **transmutation**. Transmutation can occur naturally. It can also be produced artificially by bombarding the nuclei of a substance with various particles from a particle accelerator, such as the cyclotron.

It is important to note that in all nuclear transformations (often shown as nuclear equations), the mass numbers and atomic numbers must be conserved. For example, when cobalt-60 undergoes beta radiation, the following equation represents the process:

$$\ce{^{60}_{27}Co} \rightarrow \ce{^{60}_{28}Ni} + \ce{^{0}_{-1}e^-}$$

Notice how the addition of the mass numbers adds to 60 on both sides of the equation. Similarly, the atomic numbers on both sides of the equation total 27.

The following uranium-radium decay series shows how a radioactive atom may change when it loses each kind of particle. Note that an atomic number is shown by a subscript ($_{92}$U) and the isotopic mass by a superscript (^{238}U). The alpha particle is represented by the Greek symbol α and the beta particle by β:

$$^{238}_{92}\text{U} \xrightarrow{-\alpha} {}^{234}_{90}\text{Th} \xrightarrow{-\beta} {}^{234}_{91}\text{Pa} \xrightarrow{-\beta} {}^{234}_{92}\text{U} \xrightarrow{-\alpha} {}^{230}_{90}\text{Th} \xrightarrow{-\alpha}$$

$$^{226}_{88}\text{Ra} \xrightarrow{-\alpha} {}^{222}_{86}\text{Rn} \xrightarrow{-\alpha} {}^{218}_{84}\text{Po} \xrightarrow{-\alpha} {}^{214}_{82}\text{Pb} \xrightarrow{-\beta} {}^{214}_{83}\text{Bi} \xrightarrow{-\beta}$$

$$^{214}_{84}\text{Po} \xrightarrow{-\alpha} {}^{210}_{82}\text{Pb} \xrightarrow{-\beta} {}^{210}_{83}\text{Bi} \xrightarrow{-\beta} {}^{210}_{84}\text{Po} \xrightarrow{-\alpha} {}^{206}_{82}\text{Pb} \text{ (stable)}$$

The changes that occur in radioactive reactions and the subatomic particles involved are summarized in Tables 2.2 and 2.3.

TABLE 2.2 Radioactive Decay and Nuclear Change

Type of Decay	Decay Particle	Particle Mass	Particle Charge	Change in Mass Number	Change in Atomic Number
Alpha decay	α	4	2+	Decreases by 4	Decreases by 2
Beta decay	β	0	1−	No change	Increases by 1
Gamma radiation	γ	0	0	No change	No change
Positron emission	β^+	0	1+	No change	Decreases by 1
Electron capture	e^-	0	1−	No change	Decreases by 1

TABLE 2.3 Nuclear Symbols for Subatomic Particles

Particle	Symbols	Nuclear Symbols
Proton	p	^1_1p or ^1_1H
Neutron	n	^1_0n
Electron	e^- or β^-	$^0_{-1}\text{e}$ or $^0_{-1}\beta$
Positron	e^+ or β^+	$^0_{+1}\text{e}$ or $^0_{+1}\beta$
Alpha particle	α	^4_2He or $^4_{+2}\alpha$
Beta particle	β or β^-	$^0_{-1}\text{e}$ or $^0_{-1}\beta$
Gamma ray	γ	$^0_0\gamma$

Nuclear Reactions

Nuclear fission reactions have been in use since the 1940s. The first atomic bombs used in 1945 were nuclear fission bombs. Since that time, many countries, including our own, have put nuclear fission power plants into use to provide a new energy source for electrical energy. Basically, a nuclear fission reaction is the splitting of a heavy nucleus into two or more lighter nuclei.

U-235 is bombarded with slow neutrons to produce Ba-139, Kr-94, or other isotopes and also three fast-moving neutrons:

$$^{235}_{92}U + ^{1}_{0}n \rightarrow ^{139}_{56}Ba + ^{94}_{36}Kr + 3^{1}_{0}n + Energy$$

A nuclear chain reaction is a reaction in which an initial step, such as the reaction above, leads to a succession of repeating steps that continues indefinitely. Nuclear chain reactions are used in nuclear reactors and nuclear bombs.

A **nuclear fusion** reaction is the combining of very light nuclei to make a heavier nucleus. Extremely high temperatures and pressures are required in order to overcome the repulsive forces of the two nuclei. Fusion has been achieved only in hydrogen bombs. Scientists are still trying to harness this reaction for domestic uses. The following examples show basically how the reactions occur.

Two deuterium atoms combining:

$$^{2}_{1}H + ^{2}_{1}H \rightarrow ^{4}_{2}He + Energy$$

The energy released in a nuclear reaction (either fission or fusion) comes from the fractional amount of mass converted into energy. Nuclear changes convert matter into energy. Energy released during nuclear reactions is much greater than the energy released during chemical reactions.

The Wave-Mechanical Model

In the early 1920s, some difficulties with the Bohr model of the atom were becoming apparent. Although Bohr used classical mechanics (which is the branch of physics that deals with the motion of bodies under the influence of forces) to calculate the orbits of the electron in the hydrogen atom, this discipline did not serve to explain the ability of the electron to stay in only certain energy levels without the loss of energy. Nor could it explain why a change of energy occurred only when an electron "jumped" from one energy level to another and why the electron could not exist in the atom at any energy level between these levels. According to Newton's laws, the kinetic energy of a body always changes smoothly and continuously, not in sudden jumps. The idea of only certain **quantized** energy levels being available in the Bohr atom was a very important one. The energy levels explained the existence of atomic spectra, described in the preceding sections.

Another difficulty with the Bohr model was that it worked well for only the hydrogen atom with its single electron. It did not work with atoms that had more electrons. A new approach to the laws governing the behavior of electrons inside the atom was needed. Such an approach was developed in the 1920s by the combined work of many scientists. Their work dealt with a more mathematical model usually referred to as **quantum mechanics** or **wave mechanics**. By this time, Albert Einstein had already proposed a relativity mechanics model to deal with the relative nature of mass as its speed approaches the speed of light. In the same manner, a quantum/wave mechanics model was now needed to fit the data of the atomic model. **Max Planck** suggested in his **quantum theory** of light that light has both particle-like properties and wavelike characteristics. In 1924, **Louis de Broglie**, a young French physicist, suggested that if light can have both wavelike and particle-like characteristics as Planck had suggested, perhaps particles, like electrons, can also have wavelike characteristics. In 1927, de Broglie's ideas were verified experimentally when investigators showed that electrons could produce diffraction patterns, a property associated with waves. Diffraction patterns are produced by waves as they pass through small holes or narrow slits.

In 1927, **Werner Heisenberg** stated what is now called the **uncertainty principle**. This principle states that it is impossible to know both the precise location and precise velocity of a subatomic particle at the same time. Heisenberg, in conjunction with the Austrian physicist Erwin Schrödinger, agreed with the de Broglie concept that the electron is bound to the nucleus in a manner similar to a standing wave. They developed the complex equations that describe the **wave-mechanical model** of the atom. The solution of these equations gives specific wave functions called **orbitals** based on the energy possessed by the electron. These are not the same as the electron orbits described in the Bohr model. The electron does not move in a circular orbit in the wave-mechanical model. Rather, the orbital is a three-dimensional region around the nucleus that indicates the probable location of an electron but gives no information about its pathway. The drawings in Figures 2.8a, 2.8b, and 2.8c that are found in the next section are probability distribution representations of where electrons in these orbitals might be found. These orbitals represent possible energy states a single electron in the hydrogen atom might possess. Application of wave mechanics to atoms other than hydrogen (those with more than one electron) describes similar orbital possibilities, but they vary in size compared to those of hydrogen.

Quantum Numbers and the Pauli Exclusion Principle

Each electron orbital of an atom may be described by a set of four **quantum numbers** in the wave-mechanical model. These numbers give the position with respect to the nucleus, the shape of the orbital, its spatial orientation, and the spin of the electron in the orbital.

Principal quantum number (n)
  1, 2, 3, 4, 5, etc.
The values of $n = 1, 2, 3, \ldots$

This number refers to the average distance of the orbital from the nucleus. Orbital 1 is closest to the nucleus and has the least energy. The numbers correspond to the orbits in the Bohr model. They are called energy levels.

Angular momentum (ℓ)
  quantum number
s, p, d, f orbitals
(in order of increasing energy)
The value of ℓ can $= 0, 1, \ldots,$
  $(n - 1)$
$\ell = 0$ indicates a spherical-shaped *s* orbital
$\ell = 1$ indicates a dumbbell-shaped *p* orbital
$\ell = 2$ indicates a five orbital orientation *d* orbital

This number refers to the shape of the orbital. The number of possible shapes is limited by the principal quantum number. The first energy level has only one possible shape, the *s* orbital, because $n = 1$ and because the limit of $\ell = (n - 1) = 0$. The second energy level has two possible shapes, *s* and *p*, as ℓ could have the possible values of 0 or 1. See Figures 2.8a, 2.8b, and 2.8c for representations of these shapes possible when ℓ equals 0, 1, or 2 (orbitals designated as *s*, *p*, and *d*, respectively).

Magnetic quantum number (m_ℓ)
$s = 1$ space-oriented orbital
$p = 3$ space-oriented orbitals
$d = 5$ space-oriented orbitals
$f = 7$ space-oriented orbitals
The value of *m* can equal
$-\ell, \ldots, 0, \ldots, +\ell.$

The drawings in Figures 2.8a and 2.8b show the *s*-orbital shape, which is a sphere, and the *p* orbitals, which have a dumbbell shape with three possible orientations on the axis shown. Figure 2.8c shows the five *d* orbital orientations which are mostly a four-lobed shape. The number of spatial orientations of orbitals is related to the magnetic quantum number. The possible orientations are listed and are dictated by the possible number of *m* values for any given ℓ value.

quantum number (m_s)
$+$ spin $-$ spin
The value of $m = +\frac{1}{2}$ or $-\frac{1}{2}$

Electrons are assigned one more quantum number called the spin quantum number. This describes the spin in either of two possible directions. Each orbital can be filled by only two electrons with opposite spins. The main significance of electron spin is explained by the postulate of Wolfgang Pauli. It states that in a given atom, no two electrons can have the same set of four quantum numbers ($n, \ell, m_\ell,$ and m_s). This is referred to as the **Pauli Exclusion Principle**. Therefore, each orbital in Figures 2.8a, 2.8b, 2.8c can hold only two electrons.

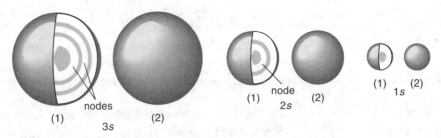

Representations of the 1s, 2s, and 3s orbitals. (1) The electron probability distribution; the nodes indicate regions of zero probability. (2) The surface that contains 90% of the total electron probability (the size of the orbital, by definition).

FIGURE 2.8a Representations of s Orbitals

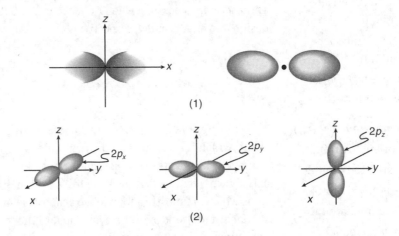

Representation of the 2p orbitals. (1) The electron probability distribution. (2) The boundary surface representations of all three orbitals.

FIGURE 2.8b Representations of p Orbitals

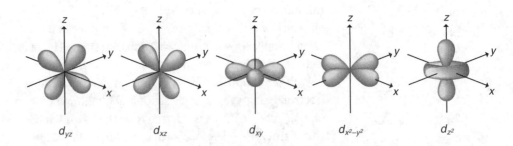

Representations of the 3d orbitals in terms of their boundary surfaces.

The subscripts of the first four orbitals indicate the planes in which the four lobes are centered.

FIGURE 2.8c Representations of d Orbitals

Quantum numbers are summarized in Table 2.4 below.

TABLE 2.4 Summary of Quantum Numbers for the First Four Levels of Energy in Any Atom

Principal Quantum No., n	Angular Momentum Quantum No., ℓ	Orbital Shape Designation	Magnetic Quantum No., m_ℓ	Number of Orbitals	Total Electrons
1	0	1s	0	1	2
2	0	2s	0	1	2
	1	2p	$-1, 0, +1$	3	6
3	0	3s	0	1	2
	1	3p	$-1, 0, +1$	3	6
	2	3d	$-2, -1, 0, +1, +2$	5	10
4	0	**4s**	0	1	2
	1	**4p**	$-1, 0, +1$	3	6
	2	**4d**	$-2, -1, 0, +1, +2$	5	10
	3	**4f**	$-3, -2, -1, 0, +1, +2, +3$	7	14

Limits of Quantum Numbers		
$n = 1, 2, 3, \ldots$	$\ell = 0, 1, \ldots, (n-1)$	$m_\ell = -\ell, \ldots, 0, \ldots, +\ell$

Energy Sublevels

The existence of the angular momentum quantum number, ℓ (sometimes called the secondary quantum number), indicates that the principal energy level, n, actually has sublevels within it. These "levels within a level" are of equal energy concerning the electron in the hydrogen atom but display slightly different energies for electrons in atoms with more than one electron. In other words, in atoms other than hydrogen, the sublevel designated by $\ell = 0$ (an s-type sublevel) in the second energy level (where $n = 2$) does not represent the same energy as the p sublevel, designated by $\ell = 1$, in that principal energy level. This means that the s orbital in the s sublevel in the second energy level is different in energy than a p orbital in the p sublevel in the second energy level. The three p orbitals in the p sublevel do represent the same energy, though, as they are in the same sublevel. Orbitals of the same energy are said to be **degenerate**. Orbitals of different energies lack degeneracy. The number of sublevels in a given principal energy level is given by the n value itself. For example, the number of sublevels in energy level $n = 4$ is 4. That's because there are 4 possible values for ℓ (0, 1, 2 and 3 since ℓ values range from 0 up to $n - 1$). The total number of orbitals in a given energy level is given by n^2. That means, for example, that energy level $n = 4$ has 16 orbitals. Some of these orbitals are degenerate (if in the same sublevel), while others lack degeneracy (if in different sublevels). The relative energies used for the filling of sublevels and orbitals associated with them are shown in Figure 2.9 for the first seven principal energy levels.

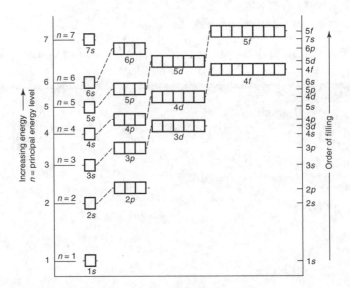

FIGURE 2.9 Approximate Relative Energy of Sublevels and Orbitals Used to Determine Filling Order

Electron Configuration

The orbitals associated with electrons in atoms are often shown when the electrons possess their lowest possible energies. Using a Bohr model term, these electrons are in their *ground state*. The ground state configuration for the electron in the hydrogen atom is designated $1s^1$. This *configuration notation* indicates that the electron is in the first energy level with a spherical probability distribution. The superscript of 1 indicates there is just one electron possessing this energy state, which has to be the case with hydrogen because it has only one electron. This electron could get "excited" to the second energy level into a spherical probability distribution (a $2s$ orbital) or a dumbbell-shaped probability distribution (a $2p$ orbital) along one of the axes in 3-D space (x, y, or z). In hydrogen, the one $2s$ and three $2p$ orbitals are degenerate. In other words, there are four degenerate orbitals on the second energy level in hydrogen. The electron in hydrogen can be associated with any of these four orbitals and possess the same energy. One *excited* configuration may be designated $2s^1$ while another may be $2p_x^{\,1}$.

For atoms with multiple electrons, the lack of degeneracy between sublevels creates an orbital filling order when ground-state electron configurations are written. For example, on a given energy level, the s sublevel is lower in energy than the p sublevel when the p sublevel is available (beginning with $n = 2$). Likewise, p is lower than d, and d is lower than f. Figure 2.9 displays the relative energies of the orbitals that are typically occupied in ground-state electron configurations, but a simplified way to remember these relative energies is given below. In Figure 2.10, the sublevels available in a given energy level (s through f) are arranged in rows and columns for energy levels 1 through 7. Diagonal arrows, starting from the top, chart the energy rankings of the sublevels, from lowest to highest.

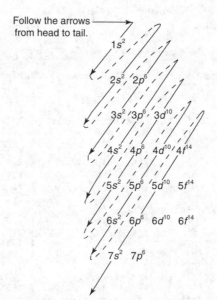

Follow the arrows from head to tail.

FIGURE 2.10 Scheme for Filling Order

Follow the arrows from tail to head and then to the tail of the next arrow to determine the electron filling order. In this way, you get the same order of filling as seen in Figure 2.9:

$$1s^2\ 2s^2\ 2p^6\ 3s^2\ 3p^6\ 4s^2\ 3d^{10}\ 4p^6\ 5s^2\ 4d^{10}\ 5p^6\ 6s^2\ 4f^{14}\ 5d^{10}\ 6p^6\ 7s^2 \ldots$$

This is required knowledge when assigning orbitals to the electrons in an atom.

Several rules must be adhered to when using the filling order correctly.

- *The Aufbau Principle.* As mandated when writing *ground-state* electron configurations, electrons should occupy the lowest energy orbitals available.

- *The Pauli Exclusion Principle.* Only two electrons can occupy any given orbital. This is because no two electrons can have the same set of quantum numbers. Since there are only two possible values for m_s, any given orbital (identified by a specific n, ℓ, and m_ℓ value) can have only two associated electrons.

- *Hund's Rule.* When filling a set of degenerate orbitals, electrons should be assigned to each orbital individually, with the same spin, before pairing them up. This rule is not noticeable when writing a configuration notation. As seen before, configuration notation only describes the principal energy level by a number, the sublevel by a letter, and the number of electrons in that sublevel by a superscript. Hund's Rule *is* noticeable, however, when using *orbital notation*.

Orbital notation is not as compact as configuration notation but can give more information. In orbital notation, each orbital is represented by a box ($\square$). Arrows are placed into the boxes to represent the electrons, with the direction of the arrow indicating the spin of the electron. Only two possible spins are possible, as there are only two possible m_s values, so the arrows can be directed only up or down. Table 2.5 shows orbital and configuration notations for the first ten elements. A careful look at the orbital notations for boron through neon shows Hund's Rule at work.

TABLE 2.5 Orbital Configuration Notations

Chemical Symbol	Atomic No.	Orbital Notation			Configuration Notation
		1s	2s	2p	
H	1	⊞	☐	☐☐☐	$1s^1$
He	2	⊞	☐	☐☐☐	$1s^2$
Li	3	⊞	⊞	☐☐☐	$1s^2 2s^1$
Be	4	⊞	⊞	☐☐☐	$1s^2 2s^2$
B	5	⊞	⊞	⊞☐☐	$1s^2 2s^2 2p^1$
C	6	⊞	⊞	⊞⊞☐	$1s^2 2s^2 2p^2$
N	7	⊞	⊞	⊞⊞⊞	$1s^2 2s^2 2p^3$
O	8	⊞	⊞	⊞⊞⊞	$1s^2 2s^2 2p^4$
F	9	⊞	⊞	⊞⊞⊞	$1s^2 2s^2 2p^5$
Ne	10	⊞	⊞	⊞⊞⊞	$1s^2 2s^2 2p^6$

Maximum electrons in orbitals at a particular sublevel:

$s = 2$ (one orbital)

$p = 6$ (three orbitals)

$d = 10$ (five orbitals)

$f = 14$ (seven orbitals)

It is worth noting that the elements potassium, calcium, and scandium, whose configuration notations are shown below, provide an example of how important it is to know the filling order as the $4s$ orbital get filled up before the $3d$.

$_{19}$K $1s^2 \, 2s^2 \, 2p^6 \, 3s^2 \, 3p^6 \, 4s^1$

$_{20}$Ca $1s^2 \, 2s^2 \, 2p^6 \, 3s^2 \, 3p^6 \, 4s^2$

$_{21}$Sc $1s^2 \, 2s^2 \, 2p^6 \, 3s^2 \, 3p^6 \, 4s^2 \, 3d^1$ (note $4s$ filled before $3d$)

In addition to knowing the proper filling order, it is important to know that the filling order is sometimes, but not often, violated. This is because there are times when a different order provides a more stable situation. This is often due to an application of Hund's Rule when there is a closeness in the energy of an s orbital in a lower energy level to d orbitals in the next higher energy level. Since the orbitals are "essentially" degenerate they fill with individual electrons before filling up completely. This often results in half-filled and completely filled sublevels, which are often associated with stability. So at atomic number 24, the $3d$ sublevel becomes half-filled by taking a $4s$ electron:

$_{24}$Cr $1s^2 \, 2s^2 \, 2p^6 \, 3s^2 \, 3p^6 \, 3d^5 \, 4s^1$

At atomic number 29, the $3d$ sublevel becomes filled by taking a $4s$ electron:

$_{29}$Cu $1s^2 \, 2s^2 \, 2p^6 \, 3s^2 \, 3p^6 \, 3d^{10} \, 4s^1$

Table 2.6 shows the electron configurations of the elements. A triangular mark indicates an outer-level electron dropping back to a lower unfilled orbital. These phenomena are exceptions to the Aufbau Principle. By following the atomic numbers throughout this chart, you will get the same order of filling as shown in Figure 2.9.

TABLE 2.6 Electron Configuration of the Elements

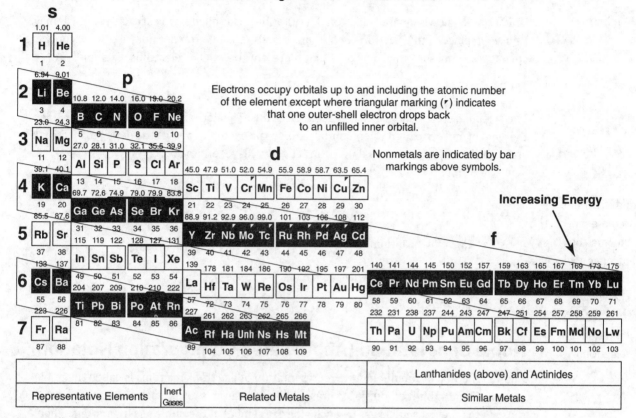

Note: Follow the order of the atomic numbers to ascertain the filling order.

By following the atomic numbers in numerical order in Table 2.6, you can plot the order of filling of the orbitals for every element shown.

Valence Electrons

Knowledge of the electrons' energy levels in an atom is crucial chemical information. The electrons in the highest energy level are those that typically dictate the chemical behavior of an element. As such, these electrons are always the outermost energy level's *s* and *p* electrons because the *d* and *f* electrons are filled belatedly. A quick look at the filling order discussed previously attests to this. Recall, as an example, how in potassium, the 4*s* orbital had an electron placed in it before the 3*d*. The outermost electrons in an atom are referred to as the **valence** electrons. Valence is a chemical term that historically refers to the combining capacity of an element, and that is what **valence electrons** prescribe. Those electrons not in the outermost energy level are referred to as the **core electrons** and generally are not directly involved in the chemical behavior of the atom.

Lewis Structures (Electron Dot Notation)

In 1916, **G. N. Lewis** devised electron dot notation, also called **Lewis structure**, which may be used in place of the electron configuration notation. Electron dot notation shows only the chemical symbol surrounded by dots that represent the valence electrons. Examples are:

$$\text{K}, \quad \cdot\text{As}\cdot, \quad \text{Sr}, \quad \ddot{:}\text{I}\ddot{:}, \quad \text{and} \quad \ddot{:}\text{Rn}\ddot{:}$$

The symbol denotes the nucleus and core electrons. The dots are arranged at the four sides of the symbol and are paired when appropriate. In the examples above, the depicted electrons are the valence electrons found in the outer energy level orbitals.

$4s^1$ is shown for potassium (K).

$4s^2\,4p^3$ are shown for arsenic (As).

$5s^2$ is shown for strontium (Sr).

$5s^2\,5p^5$ are shown for iodine (I).

$6s^2\,6p^6$ are shown for radon (Rn).

Noble Gas Notation (Abbreviated Configuration Notation)

Another method of simplifying the representation of electrons in atoms is called the noble gas notation. In this method, you represent all of the atom's lower energy electrons up to those in the previous noble gas by enclosing the symbol of the noble gas in brackets. Then the remaining electrons are represented in the usual way. Take sodium, for example, whose previous noble gas is neon. You first write [Ne] to represent sodium's inner electron structure, which is $1s^2\,2s^2\,2p^6$. This is followed by the normal notation that describes the remaining electron. Therefore, [Ne] $3s^1$ is called sodium's noble gas notation. Table 2.7 shows the noble gas notations of some of the transition elements in the fourth period of elements. Notice that the base structure of argon is used and is represented as [Ar].

Transition Elements

The elements whose *d* sublevel fills with electrons after two electrons are in the *s* sublevel of the next principal energy level are often referred to as the **transition elements**. The first examples of these are the elements between calcium, atomic number 20, and gallium, atomic number 31. Their electron configurations are the same in the 1*s*, 2*s*, 2*p*, 3*s*, and 3*p* sublevels. It is the filling of the 3*d* and changes in the 4*s* sublevels that are of interest, as shown in Table 2.7.

TABLE 2.7 Electron Configuration of Some Elements in the Fourth Period

Element	Symbol	Atomic No.	Electron Configuration							Noble Gas Notation
			$1s^2$	$2s^2$	$2p^6$	$3s^2$	$3p^6$	$3d$	$4s$	
Scandium	Sc	21						1	2	[Ar] $4s^2\, 3d^1$
Titanium	Ti	22	All					2	2	[Ar] $4s^2\, 3d^2$
Vanadium	V	23						3	2	[Ar] $4s^2\, 3d^3$
Chromium	Cr	24			the			5	1*	[Ar] $4s^1\, 3d^5$
Manganese	Mn	25						5	2	[Ar] $4s^2\, 3d^5$
Iron	Fe	26					same	6	2	[Ar] $4s^2\, 3d^6$
Cobalt	Co	27						7	2	[Ar] $4s^2\, 3d^7$
Nickel	Ni	28						8	2	[Ar] $4s^2\, 3d^8$
Copper	Cu	29						10	1*	[Ar] $4s^1\, 3d^{10}$
Zinc	Zn	30						10	2	[Ar] $4s^2\, 3d^{10}$

The asterisk (*) shows where a 4s electron is promoted into the 3d sublevel. This is because the 3d and 4s sublevels are very close in energy and because half-filled and filled sublevels have greater stability. Therefore, chromium gains stability by the movement of an electron from the 4s sublevel into the 3d sublevel to give a half-filled 3d sublevel. It then has one electron in each of the five orbitals of the 3d sublevel. In copper, the movement of one 4s electron into the 3d sublevel gives the 3d sublevel a completely filled configuration.

The fact that the electrons in the 3d and 4s sublevels are so close in energy levels leads to the possibility of some or all the 3d electrons being involved in chemical bonding. With the variable number of electrons available for bonding, it is not surprising that transition elements can exhibit variable **oxidation numbers**.

The transition elements in the other periods of the table show this same type of anomaly. They all have d sublevels that fill in the same manner.

Transition elements have several common characteristic properties:

- They often form colored compounds.
- They can have a variety of oxidation states.
- At least one of their compounds has an incomplete d electron subshell.
- They are often good catalysts.
- They are silvery blue at room temperature (except copper and gold).
- They are solids at room temperature (except mercury).
- They form complex ions.
- They are often paramagnetic due to unpaired electrons.

Periodic Table of the Elements

The development of a systematic pattern for describing the elements began near the middle of the nineteenth century. It includes the work of a number of scientists, such as Johann Dobereiner and John Newlands. Dobereiner showed that certain elements belonged in groups of three based on their similar chemical and physical behaviors. He called these groups triads. Newlands went further and showed how, if ranked according to atomic mass, elements exhibited a pattern of similar behavior every eighth element. He called these groups octaves.

Dimitri I. Mendeleev in 1869 proposed a table containing 17 columns. He is usually given credit for the first Periodic Table since he arranged elements in groups according to their atomic weights and properties. It is interesting to note that Lothar Meyer proposed a similar arrangement about the same time. Where atomic weight placement disagreed with the properties that should occur in a particular spot in the table, Mendeleev gave preference to the element with the correct properties. He even predicted elements for places that were not yet occupied in the table. These predictions proved to be amazingly accurate and led to widespread acceptance of his table.

> **TIP**
>
> Mendeleev is given credit for the first Periodic Table. It was based on placement by properties.
>
>

Periodic Law

Henry Moseley stated, after his work with X-ray spectra in the early 1900s, that the properties of elements are a periodic function of their atomic numbers. He thus changed the basis of the periodic law from atomic weight to atomic number. This is the present statement of the **periodic law**.

The Modern-Day Table

The modern-day Periodic Table has many differences from that of Mendeleev's various versions but many similarities too. The horizontal rows of the Periodic Table are called **periods** or **rows**. There are seven periods, each of which begins with an atom having only one valence electron and ends with an **inert gas** that has a complete outer shell structure. The first three periods are short, consisting of 2, 8, and 8 elements, respectively. Periods 4 and 5 are longer, with 18 elements each. Periods 6 and 7 each have 32 elements, many of which are radioactive and do not occur in nature.

In Table 2.8, note the relationship of the length of the periods to the orbital structure of the elements. In the first period, the $1s^2$ orbital is filled in the noble gas helium, He. The second period begins with the $2s^1$ orbital and ends with the filling of the $2p^6$ orbital, again in a noble gas, neon, Ne. The same pattern is repeated in the third period, which goes from $3s^1$ to $3p^6$. The eight elements from sodium, Na, to argon, Ar, complete the filling of the $n = 3$ energy level with $3s^2$ and $3p^6$. In the fourth period, the first two elements fill the $4s^2$ orbital. Beyond calcium, Ca, the pattern becomes more complicated. The next orbitals to be filled are the five $3d$ orbitals whose elements represent transition elements. Then the three $4p$ orbitals are filled, ending with the noble gas krypton, Kr. The fifth period is similar to the fourth period. The $5s^2$ orbital filling is represented by rubidium, Rb, and strontium, Sr, both of which resemble the elements directly above them on the table. Next come the transition elements that fill the five $4d$ orbitals before the next group of elements, from indium, In, to xenon, Xe, complete the three $5p$ orbitals. (Table 2.8 should be consulted for the irregularities that occur as the *d* orbitals fill.) The sixth period follows much the same pattern and has the filling order $6s^2$, $4f^{14}$, $5d^{10}$, $6p^6$. Here, again, irregularities occur and can best be followed by using Table 2.6.

The vertical columns of the Periodic Table are called **groups** or **families**. The elements in a group exhibit similar or related properties. In 1984, the IUPAC (International Union of Pure and Applied Chemistry) agreed that the groups would be numbered 1 through 18.

TABLE 2.8 Periodic Table Properties

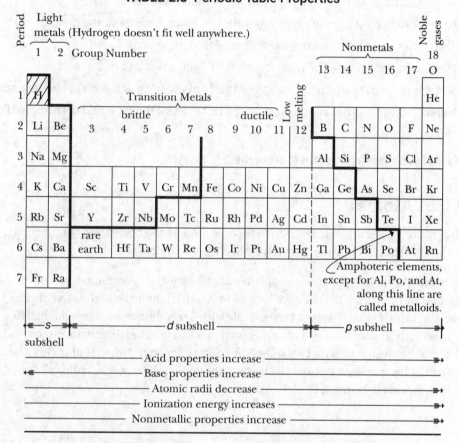

> **TIP**
> Periods are the horizontal rows 1–7. Groups are the vertical columns 1–18.

Properties Related to the Periodic Table

Metals are found on the left side of the chart (see Table 2.8) with the most active metal in the lower-left corner. Nonmetals are found on the right side with the most active nonmetal near the upper right-hand corner. The noble or inert gases are on the far right. Since the most active metals react with water to form bases, the Group 1 metals are called alkali metals. As you proceed to the right, the base-forming properties decrease and the acid-forming properties increase. The metals in the first two groups are the light metals, and those toward the center are heavy metals.

The elements found along the dark stair-step line toward the right in the Periodic Table (Table 2.8) are called **metalloids**. These elements have certain characteristics of metals and other characteristics of nonmetals. Some examples of metalloids are boron, silicon, arsenic, and tellurium. The properties of metalloids are intermediate between those of metals and those of nonmetals.

Although most metals form ionic compounds, metalloids as a group may form ionic or covalent bonds. Under certain conditions, pure metalloids conduct electricity, but do so poorly, and are thus termed **semiconductors**. This property makes the metalloids important in microcircuitry.

Here are some important general summary statements about the Periodic Table:

- Acid-forming properties increase from left to right on the table.
- Base-forming properties are high on the left side and decrease to the right.
- The atomic radii of elements decrease from left to right across a period.
- First ionization energies increase from left to right across a period.
- Metallic properties are greatest on the left side of the table and decrease to the right.
- Nonmetallic properties are greatest on the right side of the table and decrease to the left.

Table 2.8 summarizes many of these properties.

Radii of Atoms

The size of an atom is difficult to describe because atoms have no definite outer boundary. Unlike a volleyball, an atom does not have a definite circumference.

To overcome this problem, the size of an atom is estimated by describing its radius. In metals, this is done by measuring the distance between two nuclei in the solid state and dividing this distance by 2. Such measurements can be made with X-ray diffraction. For a nonmetallic element that exists in pure form as a molecule, such as chlorine, measurements can be made of the distance between nuclei for two atoms covalently bonded together. Half of this distance is referred to as the **covalent radius**. The method for finding the covalent radius of the chlorine atom is illustrated in Figure 2.11.

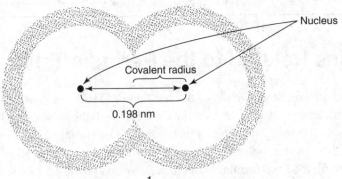

Covalent radius of Cl = $\frac{1}{2}$ (0.198) nm = 0.099 nm

FIGURE 2.11 The Covalent Radius of Chlorine

Atomic Radii in Periods

Since the number of electrons in the outer principal energy level increases as you go from left to right in each period, the corresponding increase in the nuclear charge because of the additional protons pulls the electrons more tightly around the nucleus. This attraction more than balances the repulsion between the added electrons and the other electrons, and the radius is generally reduced. The inert gas at the end of the period has a slight increase in radius because of the electron repulsion in the filled outer principal energy level.

Atomic Radii in Groups

For a group of elements, the atoms of each successive member have another outer principal energy level in the electron configuration, and the electrons there are held less tightly by the nucleus. This is so because of their increased distance from the nuclear positive charge and the shielding of this positive charge by all the core electrons. Therefore, the atomic radius increases down a group.

Ionic Radius Compared with Atomic Radius

Metals tend to lose electrons when forming positive ions. As a result, the new outermost electrons will generally exist a full energy level lower than in the atomic state. These electrons possess lower energy and have fewer core electrons shielding the pull from the nucleus and will consequently be closer to the nucleus. This makes the ions of metals smaller in size than their atomic counterparts.

Nonmetals tend to gain electrons when forming negative ions. These electrons will be added to orbitals in which electrons already exist, creating electron-electron repulsions, puffing up the orbitals, and making the ions of nonmetals larger that than their atomic counterparts.

Electronegativity

The **electronegativity** of an element is a measure of the relative strength with which the atoms of an element attract valence electrons in a chemical bond. This electronegativity number is based on an arbitrary scale ranging from 0 to 4. In general, a value of less than 2 indicates a metal.

Electronegativity decreases down a group and increases across a period. The inert gases can be ignored. The lower the electronegativity value, the more electropositive an element is said to be. The most electronegative element is near the upper-right corner. It is F, fluorine. The most electropositive element is in the lower-left corner of the chart. It is Fr, francium.

Ionization Energy

Atoms hold their valence electrons with varying levels of effective attractive force to the nucleus. This effective attractive force is determined by the number of protons in the nucleus and the amount of electron-electron repulsive effects found in the atom. If enough energy is supplied to one outer electron to remove it from its atom, this amount of energy is called the **first ionization energy**. With the first electron gone, the removal of succeeding electrons becomes more difficult because of the loss of repulsive effects that were present with a greater number of electrons. The measured values in Table 2.9 confirm this for several elements in different groups. Take note that elements in Group 1 have such large second ionization energies that the removal of a second electron is unlikely. That's why elements in this group tend to form ions with a charge of +1. Elements in Group 2 have relatively small first and second ionization energies but very large third ionization energies. This explains why elements in

Group 2 tend to form ions with a charge of +2. It should be noted that the lowest ionization energies are found with the least electronegative elements.

TABLE 2.9 Successive Ionization Energies for Selected Elements

	Element	Atomic Number	First Ionization Energy	Second Ionization Energy	Third Ionization Energy
Group 1 or IA {	Li	3	520	7,298	
	K	19	418	3,052	
Group 2 or IIA {	Be	4	899	1,757	14,849
	Mg	12	737	1,450	7,733

Sample Ionization Energies for Second and Third Electron Removal (kJ/mol)

First ionization energies, often symbolized as IE_1, can be plotted against atomic numbers, as shown in Figure 2.12 below. Follow this discussion on the graph to help you understand the peaks and valleys. Not surprisingly, the highest peaks on the graph occur for the ionization energy needed to remove the first electron from the outer energy level of the noble gases, He, Ne, Ar, Kr, Xe, and Rn. This is because as more energy levels are added to the electron structure as the atomic number increases, the additional negative fields associated with the additional electrons screen out some of the positive attraction of the nucleus. Within a period such as that from Li to Ne, the first ionization energy generally increases. This is because, as you move across a period, an increase in the number of protons (most of the time) causes the effective attraction to the outermost electron to increase despite the simultaneous addition of electrons and repulsive effects. The increased repulsive effects do outweigh the addition of the proton in a few instances, causing a drop in IE_1 as seen in the graph below going from lithium to neon or from potassium to argon.

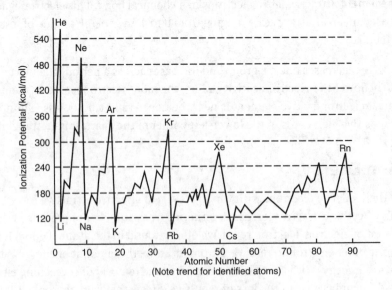

FIGURE 2.12 Ionization Energies

Bonding

Some elements show no tendency to combine with either like atoms or other kinds of elements. These elements are said to be monatomic molecules; three examples are helium, neon, and argon. A **molecule** is defined as the smallest particle of an element or a compound that retains the characteristics of the original substance. Water is a triatomic molecule since two hydrogen atoms and one oxygen atom must combine to form the substance water with its characteristic properties. When atoms do combine to form molecules, there is a shifting of valence electrons, that is, the electrons in the outer energy level of each atom. Usually, this results in the completion of the outer energy level of each atom followed by their mutual attraction. This is called a **chemical bond**. When a chemical bond forms, energy is released; when this bond is broken, energy is absorbed.

This relationship of bonding and the valence electrons of atoms can be further explained by studying the electron structures of the atoms involved. As already mentioned, the noble gases are monatomic molecules. The reason can be seen in the electron distributions of these noble gases as shown in Table 3.1.

TABLE 3.1 Noble Gas Electron Configurations

Noble Gas	Electron Distribution	Electrons in Valence Energy Level
Helium	$1s^2$	2
Neon	$1s^2\,2s^2\,2p^6$	8
Argon	$1s^2\,2s^2\,2p^6\,3s^2\,3p^6$	8
Krypton	$1s^2\,2s^2\,2p^6\,3s^2\,3p^6\,3d^{10}\,4s^2\,4p^6$	8
Xenon	$1s^2\,2s^2\,2p^6\,3s^2\,3p^6\,3d^{10}\,4s^2\,4p^6\,4d^{10}\,5s^2\,5p^6$	8
Radon	$1s^2\,2s^2\,2p^6\,3s^2\,3p^6\,3d^{10}\,4s^2\,4p^6\,4d^{10}\,4f^{14}\,5s^2\,5p^6\,5d^{10}\,6s^2\,6p^6$	8

> **TIP**
> Notice the recurrence of the octet (8) of electrons in noble gases.
>
>

The distinguishing factor in these very stable configurations, i.e., one that would take a significant amount of energy to change, is the arrangement of two *s* electrons and six *p* electrons in the valence energy level in five of the six atoms. (Note that helium, He, has only a single *s* valence energy level, which is filled with two electrons, making He a very stable atom.) This arrangement is called a **stable octet**. All other elements, those other than the noble gases, have one to seven electrons in their outer energy levels. These elements are reactive to varying degrees. When they do react to form chemical bonds, usually the electrons shift in such a way that stable octets form. In other words, in bond formation, atoms usually attain the stable electron structure of one of the noble gases. The type of bond formed is directly related to whether this structure is achieved by gaining, losing, or sharing electrons.

Types of Chemical Bonds

Chemical bonds are distinguished from other particle interactions in that the combination of atoms occurs in specific ratios with the resulting compounds exhibiting a formula. These types of interactions can occur in two different ways from a process perspective.

Ionic Bonds

> **TIP**
> An electronegativity difference between atoms of 1.7 or greater will essentially form an ionic bond.
>
>

When atoms with large electronegativity differences (generally greater than 1.6) approach each other, the interaction provides an opportunity for the particles to undergo an electron transfer and exist at a lower energy state by binding to each other instead of remaining in their isolated states. This typically occurs when metal and nonmetal atoms are around each other. Recall that metals have relatively low ionization energies, meaning they tend to lose electrons and become **cations**. On the other hand, nonmetals have a relatively strong desire to gain electrons and become **anions**. Therefore, a combination of a metal and a nonmetal encourages the transfer of electrons from one atom to another, resulting in positively and negatively charged particles that want to cling to each other. Since the charged particles that form are referred to as ions, the interaction between them creates an **ionic bond**. These ions do not retain the properties of the original atoms.

These ions do not form an individual molecule in the solid phase but are arranged into a crystal lattice or giant ion molecule containing many such ions. The simplest ratio of the ions in this aggregation is referred to as the **formula unit** and is generally what a particle of an ionic compound is called.

Covalent Bonds

When atoms with smaller electronegativity differences (generally less than 1.7) approach each other, the interaction provides an opportunity for the particles to share electrons and exist at a lower energy state by binding to each other instead of remaining isolated. This typically occurs when two nonmetal atoms are around each other. This will not occur when two metal atoms are around each other as neither has a strong desire for more electrons. On the other hand, nonmetals do have a relatively strong desire for more electrons. The only way for that to be satisfied is if the interacting atoms share their valence electrons. The difference between the level of potential energy before the bond and that once the bond has been formed is called the **bond energy**. The distance between the nuclei of the atoms in their optimum position is referred to as the **bond length**. When nonmetal atoms interact in this manner, it is called a **covalent bond**.

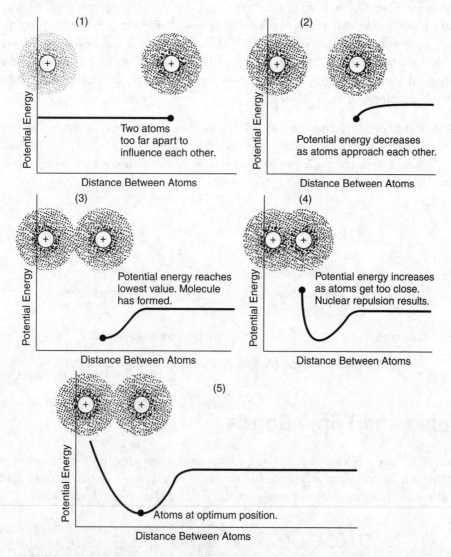

**FIGURE 3.1 Interaction of Nonmetal Atoms to Share Electrons and Form
a Covalent Bond**

When the electronegativity difference between two or more atoms is 0 or very small (not greater than about 0.4), the atoms tend to share the valence electrons in their respective outer energy levels equally. This interaction is called a **nonpolar covalent bond** as the electrons are not really pulled toward one side of the bond or another.

When the electronegativity difference is between 0.4 and 1.6, there will not be an equal sharing of electrons between the atoms involved. The shared electrons will be more strongly attracted to the atom with greater electronegativity. As the difference in the electronegativities of the two elements increases above 0.4, the polarity or degree of ionic character increases. At a difference of 1.7 or greater, the bond is 50% or more ionic character. However, when the difference is between 0.4 and 1.6, the bond is called a **polar covalent bond**.

Another way to display a shared pair of electrons in the representation of a molecule is to use a line (—) connecting the elemental symbols in place of the two dots. The line is interpreted to be a covalent bond. If the bond is polar, it can be indicated by appropriately placing symbols for partial charges at the ends of the line. The end of the bond closer to the more electronegative atom would have the symbol $\delta-$ by it as the negative electrons are drawn closer but not transferred to that atom. Conversely, the end of the bond closer to the less electronegative element would have the symbol $\delta+$ by it as the electrons are drawn away from but not completely transferred from that atom.

In the **covalent** bonds described so far, each atom contributed one of the shared electrons. In some cases, however, both electrons for the shared pair are supplied by only one of the atoms. A pair of electrons shared in this way is referred to as a coordinate covalent bond. Two examples are the bonds in NH_4^+ and H_2SO_4 as shown in Figure 3.2.

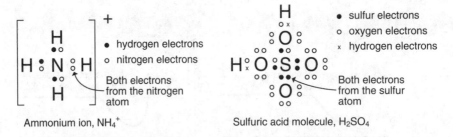

Ammonium ion, NH₄⁺   Sulfuric acid molecule, H₂SO₄

FIGURE 3.2 Examples of Coordinate Covalent Bonding

Double and Triple Bonds

To achieve a stable octet, which is an outer energy level resembling the noble gas configuration of eight electrons, some atoms must share two or even three pairs of electrons. Sharing two pairs of electrons produces a **double bond**. An example is seen in Figure 3.3.

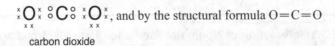

carbon dioxide

FIGURE 3.3 The Double Bonds in Carbon Dioxide

In a structural formula, only the shared pairs of electrons are shown and each pair is indicated by a short line. Therefore, a double bond is shown as two lines. The sharing of three electron pairs results in a **triple bond**. An example is seen in Figure 3.4.

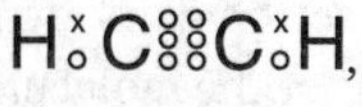

$$H \overset{x}{_\circ} C \overset{\circ\circ}{_{\circ\circ}} C \overset{x}{_\circ} H$$, and by the structural formula $H-C \equiv C-H$

acetylene

FIGURE 3.4 The Triple Bond in Acetylene

It can be assumed from these structures that there is a greater electron density between the nuclei involved as the number of bonds increases and hence a greater attractive force between the nuclei and the shared electrons. Experimental data verify that greater energy is required to break double bonds than single bonds and to break triple bonds than double bonds. Also, since these stronger bonds tend to pull atoms closer together, the atoms joined by double and triple bonds have shorter bond lengths to go along with their higher bond strengths.

Resonance Structures

It is not always possible to represent the bonding structure of a molecule by one Lewis structure. This occurs because data about the bond length and bond strength are between possible drawing configurations and really indicate a hybrid condition. To represent this situation, all possible alternatives are drawn with arrows between them. The concept that no one Lewis structure adequately represents the bonding situation in a molecule, while all of them together do, is called **resonance**. Classic examples are sulfur trioxide (SO_3) and benzene shown in Figures 3.5 and 3.6.

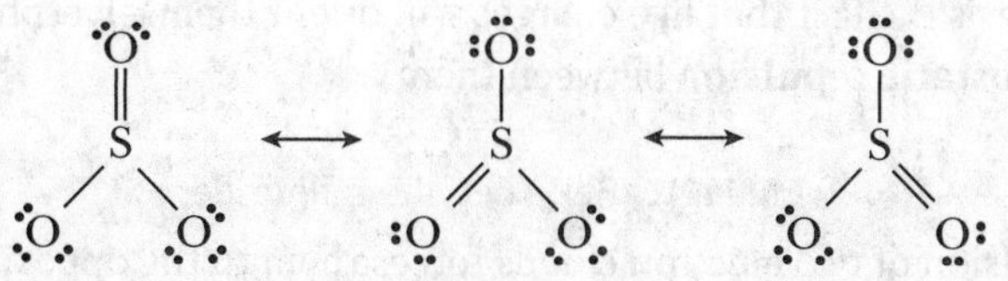

FIGURE 3.5 Resonance Structures of Sulfur Trioxide

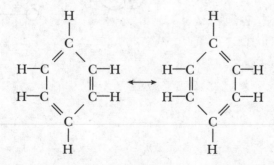

FIGURE 3.6 Resonance Structures of Benzene

In the molecule SO_3, all the bonds are known through experimentation to be similar in length and strength. This is not indicated by any of the individual structures shown above. In SO_3, no bond actually has either a single or a double bond character. Instead, the bonds in sulfur trioxide should be viewed as having a character of $1\frac{1}{3}$ bonds each. This can be determined by analyzing the resonance structures, where a total of four electron pairs are in any one location over the three structures. Resonance does not imply that the molecule alternates between the various structures shown. Instead, resonance should be interpreted as meaning all of the structures together provide a more accurate depiction of the bonding in the molecule. Similarly, the bonding character in benzene is best described as $1\frac{1}{2}$ bonds for all the carbon atoms attached to each other.

TIP
Two theories explain molecular structure. VSEPR theory uses valence shell electron pair repulsion. Hybridization theory uses changes in the orbitals of the valence electrons.

Molecular Geometry—VSEPR and Hybridization

Properties of molecules depend not only on the bonding of atoms but also on the **molecular geometry**—the three-dimensional arrangement of the molecule's atoms in space. The chemical formula reveals little information about a molecule's geometry. Only after performing many tests designed to reveal the shapes of the various molecules have chemists developed successful theories to explain certain aspects of their findings. One theory structurally accounts for molecular bond angles. The other is used to describe changes in the orbitals that contain the valence electrons of a molecule's atoms. The structural theory that deals with the bond angles is called the **VSEPR theory**, whereas the one that describes changes in the orbitals that contain the valence electrons is called the **hybridization theory**. (VSEPR represents Valence Shell Electron Pair Repulsion.)

VSEPR—Electrostatic Repulsion

VSEPR uses as its basis the fact that like charges will orient themselves in such a way as to diminish the **electrostatic repulsion** between them.

EXAMPLE: BeF_2, beryllium fluoride

1. Mutual repulsion of two electron clouds forces them to the opposite sides of a sphere. This is called a **linear arrangement**.

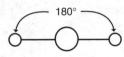

2. Minimum repulsion between three electron pairs occurs when the pairs are at the vertices of an equilateral triangle inscribed in a sphere. This is called a **trigonal-planar arrangement**.

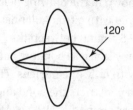

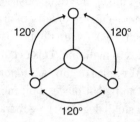

3. Four electron pairs are farthest apart at the vertices of a tetrahedron inscribed in a sphere. This is called a **tetrahedral arrangement** or a tetrahedral-shaped distribution of electron pairs.

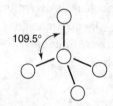

4. Mutual repulsion of six identical electron clouds directs them to the corners of an inscribed regular octahedron. This is called an **octahedral arrangement**.

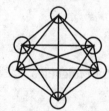
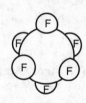
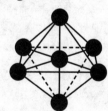

> **TIP**
> These basic arrangements are important to learn!
>
>

VSEPR and Unshared Electron Pairs

Ammonia (NH_3) and water (H_2O) are examples of molecules in which the central atom has both shared and unshared electron pairs. Here is how the VSEPR theory accounts for the geometries of these molecules.

The Lewis structure of ammonia shows that in addition to the three electron pairs the central nitrogen atom shares with the three hydrogen atoms, it also has one unshared pair of electrons:

$$H : \overset{\cdot\cdot}{\underset{\overset{|}{H}}{N}} : H$$

VSEPR theory postulates that the lone pair occupies space around the nitrogen atom just as the bonding pairs do. Thus, as in the methane molecule shown in an example in the preceding section, the electron pairs maximize their separation by assuming the corners of a tetrahedron. Lone pairs do occupy space, but our description of the observed shape of a molecule refers to the positions of atoms only. Consequently, as shown in the drawing below, the molecular geometry of an ammonia molecule is that of a pyramid with a triangular base. The angle between the bonds is shown to be a bit smaller than 109.5° found between the bonds in methane. This occurs because not only do lone pairs occupy space as bonding pairs do, they also occupy a bit *more* space as lone pairs are associated with only one nucleus. This spreading out places a bit more repulsion on the bonding pairs and constricts the bond angle by a small degree. The general VSEPR formula for a molecule such as ammonia (NH_3) is AB_3E, where A replaces N, B replaces H, and E represents the unshared electron pair.

A water molecule has two unshared electron pairs and can be represented as an AB_2E_2 molecule. Here, the oxygen atom is at the center of a tetrahedron, with two corners occupied by hydrogen atoms and two by the unshared electron pairs, as shown below. Again, VSEPR theory states that the lone pairs occupy space around the central atom but that the actual shape of the molecule is determined only by the positions of the atoms. In the case of water, this results in a "bent," or angular, molecule. An even smaller bond angle is shown in Figure 3.7 compared to ammonia, due to two lone pairs spreading out in space a bit more than the bonding pairs.

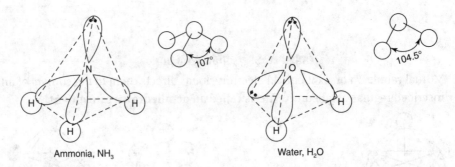

Ammonia, NH_3 Water, H_2O

FIGURE 3.7 VSEPR Theory Applied to Ammonia and Water Molecules

VSEPR and Molecular Geometry

Table 3.2 summarizes the molecular shapes associated with particular types of molecules. In VSEPR theory, double and triple bonds are treated in the same way as single bonds. It is helpful to use the Lewis structures and this table together to predict the shapes of molecules with double and triple bonds as well as the shapes of polyatomic ions.

TABLE 3.2 Summary of Molecular Shapes

Type of Molecule	Molecular Shape	Atoms Bonded to Central Atom	Lone Pairs of Electrons	Formula Example	Lewis Structure
Linear		2	0	BeF_2	
Bent		2	1	$SnCl_2$	
Trigonal-planar		3	0	BF_3	
Tetrahedral		4	0	CH_4	
Trigonal-pyramidal		3	1	NH_3	
Bent		2	2	H_2O	
Trigonal-bipyramidal		5	0	PCl_5	
Octahedral		6	0	SF_6	

The Polarity of Molecules

The combination of the molecular geometry, the polarity of the bonds, and the placement of unshared pairs of electrons all contribute to the overall polarity of a molecule. A *polar molecule* has an uneven distribution of charge, while a *nonpolar molecule* is balanced concerning the electrical charge. Molecules of ammonia and water lack symmetry with regards to their charge distribution. They are polar molecules as shown in Figure 3.8.

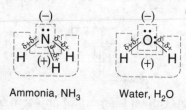

Ammonia, NH_3 Water, H_2O

FIGURE 3.8 Polar Molecules of Ammonia and Water

In ammonia, the lone pair on the top of the molecule in combination with the polar bonds make the nitrogen side of the molecule negative. The hydrogen side of the molecule is positive as it lacks lone pairs. So the hydrogen end of each bond is partially positive. Similarly in water, the oxygen side of the molecule is negative and the hydrogen side is positive.

As seen in Figure 3.9, Carbon tetrachloride (CCl_4) and boron trifluoride (BF_3) are nonpolar molecules as their charge distribution is balanced. Despite the polar bonds in both molecules and the lone pairs in BF_3, the symmetrical molecular shapes of each cancel out the polarity of the bonds.

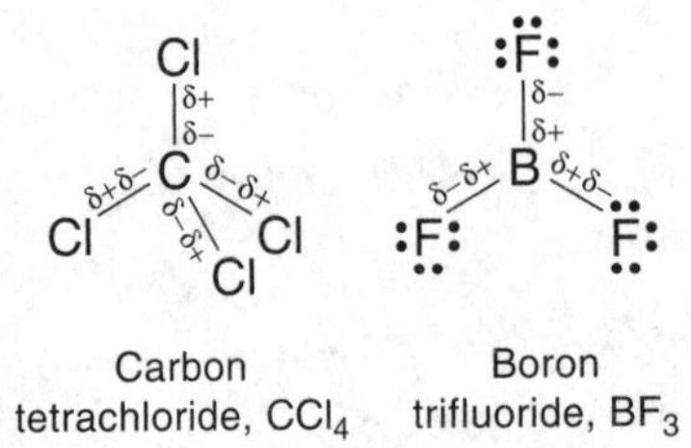

Carbon
tetrachloride, CCl_4

Boron
trifluoride, BF_3

FIGURE 3.9 Nonpolar Molecules of Carbon Tetrachloride and Boron Trifluoride

Hybridization

For many molecules, the configurations derived by VSEPR theory could not be realized unless changes in the electron orbitals of the connecting atoms took place. The valence electrons in atoms are not set up to produce the kinds of orbital overlap that results in a covalent bond at the angles we find when analyzing the molecules. Therefore, in the presence of other atoms, the electron energy states that are possible (reflected by new orbital possibilities) are changed. This modification of the types of orbitals available to the valence electrons when atoms are bonding to other atoms is called **hybridization**. Briefly stated, this means that chemists envision that two or more pure atomic orbitals (usually *s*, *p*, and *d*) can be mixed to form two or more new hybrid atomic orbitals that are identical and conform to the known shapes of molecules. Hybridization can be illustrated as follows:

1. *sp* Hybrid Orbitals

    Spectroscopic measurements of beryllium fluoride, BeF_2, reveal a bond angle of 180° and equal bond lengths as described in Figure 3.10.

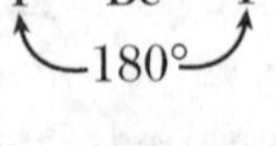

$$F\!-\!Be\!-\!F$$
$$\underbrace{\qquad\qquad}_{180°}$$

FIGURE 3.10 Bond Angles in Beryllium Fluoride

The ground-state electron structure of beryllium is shown in Figure 3.11:

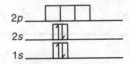

FIGURE 3.11 Orbital Diagram of an Atom of Beryllium

This configuration of valence electrons makes it seem like beryllium is not able to form two identical bonds as seen in Figure 3.10. The valence electrons are paired and not available to overlap with two other electron orbitals to form two covalent bonds. To accommodate the experimental data, we theorize that a $2s$ electron is excited to a $2p$ orbital; then the two orbitals hybridize to yield two identical orbitals called sp orbitals. Each contains one electron but is capable of holding two electrons. The process is described graphically in Figure 3.12.

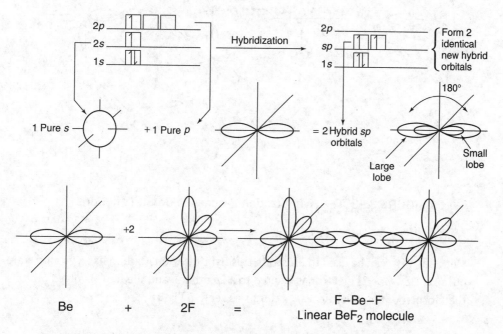

FIGURE 3.12 The Hybridization Process in Beryllium Fluoride

2. sp^2 Hybrid Orbitals

Boron trifluoride, BF_3, has bond angles of 120° and bonds of equal strength. To accommodate these data, the boron atom hybridizes from its ground state of $1s^2 2s^2 2p^1$. As seen in Figure 3.13, one $2s$ electron is excited to a $2p$ orbital. The three involved orbitals then form three new, identical sp^2 orbitals.

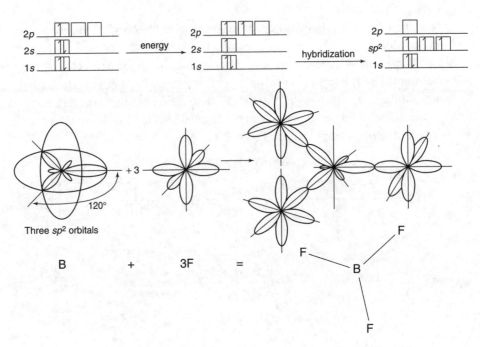

FIGURE 3.13 The Hybridization Process in Boron Trifluoride

3. *sp*³ Hybrid Orbitals

Methane, CH_4, can be used to illustrate this hybridization. Carbon has a ground state of $1s^2\,2s^2 2p^2$. One $2s$ electron is excited to a $2p$ orbital, and the four involved orbitals then form four new, identical sp^3 orbitals as seen in Figure 3.14.

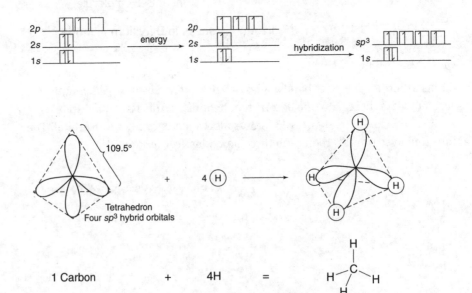

FIGURE 3.14   The Hybridization Process in Methane

In some compounds where only certain sp^3 orbitals are involved in bonding, distortion in the bond angle occurs because of increased unbonded electron repulsion as previously mentioned. Examples are shown in Figures 3.15a and 3.15b.

a.   Water, H_2O

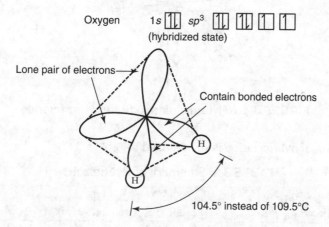

FIGURE 3.15a Hybrid Orbitals in Water Holding Both Bonding and Nonbonding (Lone Pair) Electrons

b.   Ammonia, NH_3

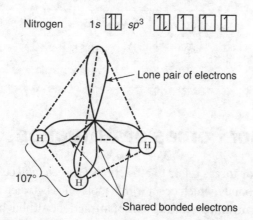

FIGURE 3.15b Hybrid Orbitals in Ammonia Holding Both Bonding and Nonbonding (Lone Pair) Electrons

4. sp^3d^2 Hybrid Orbitals

These orbitals are formed from the hybridization of an s and a p electron promoted to d orbitals and transformed into six equal $sp^3 d^2$ orbitals. The spatial form is shown in Figure 3.16. Sulfur hexafluoride, SF_6, illustrates this hybridization.

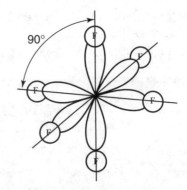

FIGURE 3.16 Hybridization in Sulfur Hexafluoride

The concept of hybridization is summarized in Table 3.3.

TABLE 3.3 Summary of Hybridization

Number of Bonds	Number of Unused Electron Pairs	Type of Hybrid Orbital	Angle Between Bonded Atoms	Geometry	Example
2	0	*sp*	180°	Linear	BeF_2
3	0	*sp²*	120°	Trigonal-planar	BF_3
4	0	*sp³*	109.5°	Tetrahedral	CH_4
3	1	*sp³*	107°	Pyramidal	NH_3
2	2	*sp³*	104.5°	Angular	H_2O
6	0	*sp³d²*	90°	Octahedral	SF_6

Intermolecular Forces of Attraction

The term **intermolecular forces** refers to attractions *between* molecules. These are distinctly different from covalent bonds, which occur *within* molecules between atoms. Covalent bonds are often referred to as **intramolecular forces** of attraction. Although it is proper to refer to all intermolecular forces as **van der Waals forces**, named after Johannes van der Waals (Netherlands), this concept should be expanded for clarity.

Dipole-Dipole Attraction

One type of van der Waals force is **dipole-dipole attraction**. It was shown in the discussion of polar molecules that the unsymmetrical distribution of electric charges leads to positive and negative charges in the molecules, which are referred to as **dipoles**. In polar molecular substances like the hydrogen chloride shown in Figure 3.17, the dipoles line up so that the positive pole of one molecule attracts the negative pole of another. This is much like the lineup of small bar magnets, although electrostatic in nature. The force of attraction between polar molecules is called dipole-dipole attraction. These attractive forces are less than those caused by the full charges carried by ions in ionic crystals.

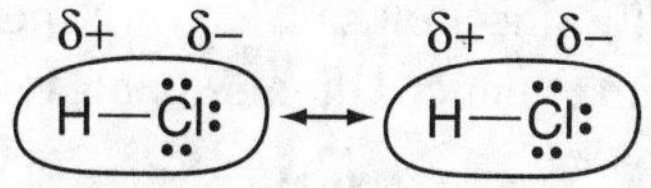

Polar hydrogen chloride molecules

FIGURE 3.17 Dipole-Dipole Attraction in Hydrogen Chloride

London Dispersion Forces

Another type of van der Waals forces is called **London dispersion forces**. Found in both polar and nonpolar molecules, London dispersion forces occur because a molecule or an atom that is usually nonpolar sometimes becomes polar because of the probabilistic nature of an electron orbital. In other words, the negatively charged electron has a chance of being in many locations. As a result, an uneven charge distribution may happen at any instant. When this occurs, the molecule/atom has a temporary dipole. This temporary dipole can then cause a second, adjacent molecule/atom to be distorted, that is, have an induced dipole. As a result, the positive nucleus of the second molecule/atom becomes attracted to the negative end of the first molecule/atom and vice versa.

London dispersion forces are about one-tenth of the force of most dipole interactions. In fact, they are the weakest of all the electrical forces that act between atoms or molecules. London dispersion forces do increase in magnitude, however, when more electron orbitals are associated with the atoms in the molecule. This is generally related to the mass of the molecule. A greater mass indicates more electrons are present, increasing the temporary polarizability of the molecule.

London dispersion forces help to explain why nonpolar substances, such as noble gases and the halogens, condense into liquids and then freeze into solids when the temperature is lowered sufficiently. In general, they also explain why liquids composed of discrete molecules with no permanent dipole attraction have low boiling points relative to their molecular masses. Additionally, compounds in the solid-state that are bound mainly by London dispersion forces have rather soft crystals, are easily deformed, and vaporize easily. Because of their low intermolecular forces, these solids have low melting points and they evaporate so easily that it may occur at room temperature. Examples of such solids are iodine crystals and mothballs (paradichlorobenzene and naphthalene).

TIP
Weakest of all, the London dispersion forces are one-tenth the force of most dipole attractions.

Hydrogen Bonds

When a hydrogen atom is bonded to a small, highly electronegative atom, the hydrogen atom's positive charge will have an enhanced attraction for neighboring electron pairs on the small, highly electronegative atom in neighboring molecules. This special kind of dipole-dipole attraction is called a **hydrogen bond**. The more strongly polar the molecule is, the more effective the hydrogen bonding is in binding the molecules into a larger unit. As a result, the boiling points of such molecules are higher than those of similar polar molecules not exhibiting hydrogen bonding. Hydrogen bonding typically occurs *between* molecules that contain hydrogen covalently bonded to nitrogen, oxygen, or fluorine *within* the molecule. These three small but highly electronegative atoms create significantly polar bonds with the hydrogen atoms in the molecule, making the molecule very polar. The highly positive hydrogen end of the molecule is very attracted to the highly negative end of another molecule of the substance (where the nitrogen, oxygen, or fluorine atoms reside with their associated nonbonding electron pairs). This process creates an enhanced dipole-dipole bond.

Studying Figure 3.18 shows that in the series of compounds consisting of H_2O, H_2S, H_2Se, and H_2Te, an unusual rise in the boiling point of H_2O occurs that is not in keeping with the typical slow increase of boiling point as molecular mass increases. Instead of the expected slope of the line between H_2O and H_2S, which is shown in Figure 3.18 as a dashed line, the actual boiling point of H_2O is quite a bit higher: 100°C. The explanation is that hydrogen bonding occurs in H_2O but not to any significant degree in the other compounds.

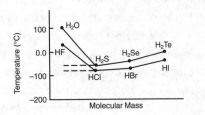
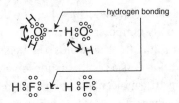

FIGURE 3.18 Boiling Points of Hydrogen Compounds with Similar Lewis Structures

This same phenomenon occurs with the hydrogen halides (HF, HCl, HBr, and HI). Note in Figure 3.18 that hydrogen fluoride, HF, which has strong hydrogen bonding, shows an unexpectedly high boiling point.

Hydrogen bonding also explains why some substances have unexpectedly low vapor pressures, high heats of vaporization, and high melting points. In order for vaporization or melting to take place, molecules must be separated. Energy must be expended to break hydrogen bonds and thus break down the larger clusters of molecules into separate molecules. As with the boiling point, the melting point of H_2O is abnormally high when compared with the melting points of the hydrogen compounds of the other elements having six valence electrons, which are chemically similar but which have no apparent hydrogen bonding.

Metallic Bonds

When only metal atoms have the opportunity to interact, neither ionic nor covalent bonding is possible. The transfer of electrons won't occur because none of the atoms have a strong desire to accept the electrons that others have a desire to give up. Likewise, the sharing of electrons won't take place because none of the atoms would like to possess more electrons. What does occur when metal atoms are around each other is that the loosely held electrons sort of envelop the positive cores of the atoms they don't have a strong desire to be around individually. In most metals, one or more of the valence electrons become detached from the atoms and migrate in a "sea" of free electrons among the positive metal ions that result from the loss of these electrons. A better way to describe the situation is that the electrons enter delocalized electron orbitals that are not associated with any particular atom but, instead, belong to the whole metal sample. The **metallic bond** can be envisioned as an attraction to the electrons in the delocalized orbitals by the positive ion cores of the metal atoms. The attractive force strength varies with the amount of positive nuclear charge and the number of electrons in the electron sea.

TIP

Notice how hydrogen bonding significantly elevates the boiling point of H_2O above the expected value.

TIP

Metallic bonds are like positive ions in a "sea" of electrons.

Properties of Differently Bonded Substances

The properties of substances vary by how the particles in the substances are bonded to each other. In particular, the relative strengths of the bonds should be compared. In general, covalent bonding is the strongest type of bond followed by ionic, metallic, and then intermolecular bonding, which is the weakest. Hydrogen bonding is the strongest type of intermolecular bond followed by dipole-dipole attraction and then London dispersion forces (LDFs). These LDFs become stronger when more electrons are present. When laboratory experiments find, in general, the properties predicted by the various bonding models, credence is given to the theory describing the bonding.

Properties of Ionic Substances

Ionic substances are those in which fully charged ions are bound by ionic bonds. Laboratory experiments reveal that, in general, ionic substances are characterized by the following properties:

1. In the solid phase at room temperature, they do not conduct appreciable electric current.

2. In the liquid phase, they are relatively good conductors of electric current.

3. They have relatively high melting and boiling points. There is a wide variation in the properties of different ionic compounds. For example, potassium iodide (KI) melts at 686°C and boils at 1,330°C, while magnesium oxide (MgO) melts at 2,800°C and boils at 3,600°C. Both KI and MgO are ionic compounds.

4. They have relatively low volatilities and low vapor pressures. In other words, they do not readily evaporate at room temperature.

5. They are brittle and easily broken when stress is exerted on them.

6. Those that are soluble in water form electrolytic solutions that are good conductors of electricity. There is, however, a wide range in the solubilities of ionic compounds. For example, at 25°C, 92 grams of sodium nitrate ($NaNO_3$) dissolve in 100 grams of water, while only 0.0002 grams of barium sulfate ($BaSO_4$) dissolve in the same mass of water.

Properties of Molecular Substances

Molecular substances include those in which molecules are bound by intermolecular forces of attraction (i.e., dipole-dipole, LDFs, or hydrogen bonding). Covalent bonds are present within the molecules (intramolecular) but do not dictate the properties of a molecular substance.

Experiments have shown that molecular substances have the following general properties:

1. Neither the liquids nor the solids conduct electric current appreciably.

2. Many exist as gases at room temperature and atmospheric pressure, and many of the solids and liquids are relatively volatile.

3. The melting points of the solid crystals are relatively low.

4. The boiling points of the liquids are relatively low.

5. The solids are generally soft and have a waxy consistency.

6. Polar molecular substances generally dissolve in water. Nonpolar molecular substances generally dissolve in nonpolar molecular solvents.

7. A large amount of energy is often required to decompose the substance chemically into simpler substances.

Properties of Metallic Substances

Metallic substances include those in which atoms are bound to each other by metallic bonding. Experiments reveal that metallic substances have the following general properties:

1. All metals, except for mercury, exist as solids at room temperature as they have relatively high melting points.

2. Metals conduct electricity in both the solid and liquid states.

3. Metals conduct heat well.

4. Metals have low vapor pressures.

5. Metals are not soluble in water.

CHAPTER 4

Chemical Formulas

Learning Objectives

In this chapter, you will learn how to:

- Recall and use the basic rules about ionic charges to write correct formulas for ionic compounds. This includes writing formulas with polyatomic ions.
- Recall and use the basic rules about writing compounds for covalent (molecular) compounds.
- Name compounds (acids, bases, and salts) using the Stock system and the prefix system, and write their formulas.
- Calculate the formula mass of a compound and the percent composition of each element.
- Calculate the empirical formula when given the percent composition of each element and how to find the true formula when given the formula mass.
- Write a simple balanced equation, indicating the phase (or state) of the reactants and products.

Naming and Writing Chemical Formulas

With the knowledge you have about atomic structure, the significance of each element's placement in the Periodic Table, and the bonding of atoms in ionic and covalent arrangements, you can now use this information to write appropriate formulas and name the resulting products. Obviously, many compounds can result. Some system of writing the names and formulas of these many combinations was needed. The system explained in this text is an organized way of accomplishing this. It uses three categories for those compounds containing only two elements:

> CATEGORY I—Binary ionic compounds where the metal present forms only a single type of positively charged ion (cation)

> CATEGORY II—Binary ionic compounds where the metal forms more than one type of ionic compound with a given negatively charged ion (anion)

> CATEGORY III—Binary covalent compounds formed between two nonmetals

Table 4.1 is a list of ions that are often encountered in a first-year chemistry course. Although using the Periodic Table can help you write the symbol and apparent charge of cations and anions, knowing these common ions can help you write formulas and equations.

TABLE 4.1 Table of Common Ions Used in a First-Year Course

Metals Cations (+)

Monovalent I		Bivalent II		Trivalent III		Tetravalent IV		V	
Hydrogen	H	Barium	Ba	Aluminum	Al	Carbon	C	Arsenic(V)*	As
Potassium	K	Calcium	Ca	Gold(III)*	Au	Silicon	Si	Phosphorus(V)*	P
Sodium	Na	Cobalt	Co	Arsenic(III)*	As	Manganese(IV)*	Mn	Antimony(V)*	Sb
Silver	Ag	Magnesium	Mg	Chromium	Cr	Tin(IV)*	Sn	Bismuth(V)*	Bi
Mercury(I)*	Hg	Lead(II)*	Pb	Iron(III)*	Fe	Platinum	Pt		
Copper(I)*	Cu	Zinc	Zn	Phosphorus(III)*	P	Sulfur	S		
Gold(I)*	Au	Mercury(II)*	Hg	Antimony(III)*	Sb	Lead(IV)*	Pb		
Ammonium†	(NH_4)	Copper(II)*	Cu	Bismuth(III)*	Bi				
		Iron(II)*	Fe						
		Manganese(II)*	Mn						
		Tin(II)*	Sn						

*Use of the Roman numeral, instead of the suffix, is now preferred; for example, iron(II) oxide instead of ferrous oxide.

†Polyatomic ion

Nonmetals‡ Anions (−)

Monovalent I		Bivalent II		Trivalent III		Tetravalent IV	
Fluorine	F	Oxygen	O	Nitrogen	N	Carbon	C
Chlorine	Cl	Sulfur	S	Phosphorus	P		
Bromine	Br						
Iodine	I						

‡Last syllable of nonmetal name is changed to -ide in binary compound.

Polyatomic Ions (−)

Monovalent I		Bivalent II		Trivalent III		Tetravalent IV	
Hydroxide	(OH)	Carbonate	(CO_3)	Borate	(BO_3)	Ferrocyanide	$[Fe(CN)_6]$
Hydrogen carbonate (or Bicarbonate)	(HCO_3)	Sulfite	(SO_3)	Phosphate	(PO_4)		
Nitrite	(NO_2)	Sulfate	(SO_4)	Phosphite	(PO_3)		
Nitrate	(NO_3)	Tetraborate	(B_4O_7)	Ferricyanide	$[Fe(CN)_6]$		
Hypochlorite	(ClO)	Silicate	(SiO_3)				
Chlorate	(ClO_3)	Chromate	(CrO_4)				
Chlorite	(ClO_2)	Oxalate	(C_2O_4)				
Perchlorate	(ClO_4)						
Acetate	$(C_2H_3O_2)$						
Permanganate	(MnO_4)						
Hydrogen sulfate (or Bisulfate)	(HSO_4)						

Category I—Binary Ionic Compounds

Category I binary ionic compounds contain metallic ions from groups 1 and 2 of the Periodic Table. These metallic ions have only one type of charge. The binary ionic compounds formed are composed of a positive ion (cation) that is written first and a negative ion (anion). The following rules show how to name and write the formulas for binary ionic compounds. $CaCl_2$ is used as an example.

1. Name the cation first and then the anion.

2. The monoatomic (one-atom) cation takes its name from the name of the element. Therefore, the calcium ion, Ca^{2+}, is called calcium and its chemical symbol appears first.

3. The monoatomic anion with which the cation combines is named by taking the root of the element's name and adding -ide. The anion's name comes second. Therefore, the chlorine ion, Cl^-, is called chloride.

4. The name of this compound is calcium chloride.

A quick way to determine the formula of a binary ionic compound is to use the **crisscross rule**.

EXAMPLE

To determine the formula for calcium chloride, first write the ionic forms with their associated charges.

Next move the numerical value of the metal ion's superscript (without the charge) to the subscript of the nonmetal's symbol. Then take the numerical value of the nonmetal's superscript and make it the subscript of the metal as shown above.

Note that the numerical value 1 is not shown in the final formula.

You now have the chlorine's 1 as the subscript of the calcium and the calcium's 2 as the subscript of the chloride. As a result, you have $CaCl_2$ as the final formula for calcium chloride.

The crisscross rule generally works very well. In one situation, though, you have to be careful. Suppose you want to write the compound formed when magnesium reacts with oxygen. Magnesium, an alkaline earth metal in group 2 forms a $2+$ cation, and oxygen forms a $2-$ anion. You would predict its formula be Mg_2O_2, but this is incorrect. After you do the crisscrossing (unless you know that the compound actually exists, like H_2O_2), you need to reduce all the subscripts by a common factor. In this example, you can divide all the subscripts by 2 to get the correct formula for magnesium oxide, MgO.

When you attempt to write a formula, you should know whether the substance actually exists. For example, you could easily write the formula for carbon nitrate, but no chemist has ever prepared this compound. Table 4.2 shows the formulas for some common Category I binary ionic compounds.

> **REMEMBER**
> Reduce all subscripts by a common factor unless you are sure the compound exists, like H_2O_2.

TABLE 4.2 Examples of Category I Binary Ionic Compounds

Ions Present*	Formula	Name
K^+, Cl^-	KCl	Potassium chloride
Na^+, I^-	NaI	Sodium iodide
Ca^{2+}, S^{2-}	CaS	Calcium sulfide
Al^{3+}, F^-	AlF_3	Aluminum fluoride
Li^+, N^{3-}	Li_3N	Lithium nitride

Ionic charges are shown as numerical exponents followed by the charge.

Category II—Binary Ionic Compounds

In Category II binary ionic compounds, the metals form more than one ion, each with a different charge. The metallic ions (cation) ionically bind with a negatively charged ion (anion).

Table 4.3 lists many of the metals that form more than one type of cation and therefore more than one binary ionic compound with a given anion.

TABLE 4.3 Common Category II Cations (Multivalent Metals)

Ion	Systematic Name	Ion	Systematic Name
Fe^{3+}	Iron(III)	Sn^{4+}	Tin(IV)
Fe^{2+}	Iron(II)	Sn^{2+}	Tin(II)
Cu^{1+}	Copper(I)	Pb^{4+}	Lead(IV)
Cu^{2+}	Copper(II)	Pb^{2+}	Lead(II)
Hg_2^{2+}	Mercury(I)*	Hg^{2+}	Mercury(II)

This form of mercury(I) ions always occurs bonded together as a Hg_2^{2+} ion.

Although the following metals are "transition" metals, they form only one type of cation. So a Roman numeral is not used when naming their compounds.

Ag^{1+} Silver

Cd^{2+} Cadmium

Zn^{2+} Zinc

> **EXAMPLE**
>
> The compound containing the Fe^{2+} ion and the compound containing the Fe^{3+} ion both combine with the chloride ion to form two different compounds. Using the crisscross system, you get the formula $FeCl_2$ for iron(II) chloride.
>
> $$Fe^{2+} \quad Cl^{1-}$$
> $$Fe \qquad Cl_2$$
>
> The compound formed using the Fe^{3+} ion and the chloride ion is $FeCl_3$, which is iron(III) chloride.
>
> $$Fe^{3+} \quad Cl^{1-}$$
> $$Fe \qquad Cl_3$$

The names iron(II) chloride and iron(III) chloride are arrived at by using the Roman numerals in parentheses to indicate the charge of the metallic ion used as the cation. Using Roman numerals this way—to indicate the charge on the ion—is called the **Stock system**.

Another, older system of naming Category II binary ionic compounds is still seen in some books. Simply stated, for metals that form only two ions, the ion with the higher charge has a name ending in -ic and the ion with the lower charge has a name ending in -ous. In this system, Fe^{3+} is called the ferric ion and Fe^{2+} is called the ferrous ion. The names for $FeCl_2$ and $FeCl_3$ are then ferric chloride and ferrous chloride, respectively.

Table 4.4 shows the formulas for some common Category II binary ionic compounds.

TABLE 4.4 Examples of Category II Binary Ionic Compounds

Formula	Name
CuCl	Copper(I) chloride
HgO*	Mercury(II) oxide
FeO*	Iron(II) oxide
MnO_2†	Manganese(IV) oxide
$PbCl_2$	Lead(II) chloride

*The subscripts are reduced and are not written because subscripts of 1 are understood.
†The subscripts are reduced.

Figure 4.1 displays a modified periodic chart that shows the location of the common Category I and Category II cations. Also shown in this chart are the common nonmetallic monatomic ions as anions.

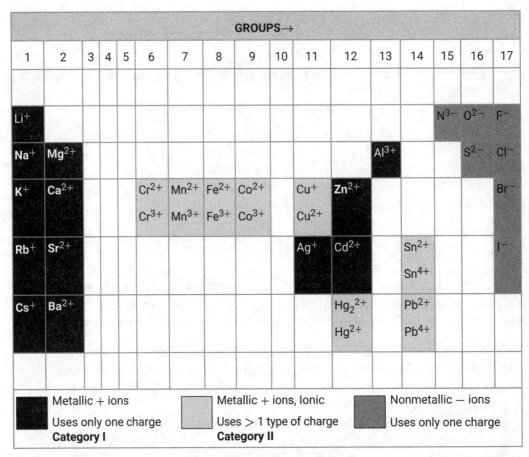

FIGURE 4.1 Summary of Common Cations (+) and Anions (−) by Category and Their Positions in the Periodic Chart

Category I and II Ionic Compounds Formed with Polyatomic Ions

Another type of ionic compound contains polyatomic ions. A **polyatomic ion** is a group of atoms that *act like a single ion* when forming a compound. The bonds within these polyatomic ions are predominantly covalent. However, the group as a whole has an excess charge, which is usually negative, because of an excess of electrons. Consequently, another way of thinking about a polyatomic ion is like a molecule with a charge. If the compounds formed with the polyatomic ions consist of three elements, they are called **ternary** compounds.

Polyatomic ions have special names and formulas that you must memorize. Table 4.1 contains the names and ionic charges of the common polyatomic ions encountered in a first-year chemistry course. Note that only one commonly used positively charged polyatomic ion is in Table 4.1, the ammonium ion, NH_4^+.

Also notice in Table 4.1 that several of the polyatomic anions contain an atom of a given element and a different number of oxygen atoms, such as NO_2 and NO_3. When there are two members of such a series, the name of the one with fewer oxygen atoms ends in -ite and the name of the one with more oxygen atoms ends in -ate. Table 4.5 shows examples of polyatomic ions of sulfur and oxygen.

TABLE 4.5 Polyatomic Ions Containing the *-ite* and *-ate* Forms of Sulfur

Ionic Formula	Name of the Ion	Sample Formula	Name of Compound
SO_3^{2-}	Sulfite	Na_2SO_3	Sodium sulfite
SO_4^{2-}	Sulfate	Na_2SO_4	Sodium sulfate

Sometimes an element combines with oxygen to form more than just two polyatomic ions, such as ClO^-, ClO_2^-, ClO_3^-, and ClO_4^-. When this occurs, the prefix *hypo-* is used to name the polyatomic ion with the fewest oxygen ions and the prefix *per-* to name the polyatomic ion with the most oxygen ions. Table 4.6 shows this series of polyatomic ions.

TABLE 4.6 Polyatomic Ions Containing Chlorine and Oxygen

Ionic Formula	Name of the Ion
ClO^-	Hypochlorite
ClO_2^-	Chlorite
ClO_3^-	Chlorate
ClO_4^-	Perchlorate

Writing Formulas for Compounds with Polyatomic Ions

When writing formulas using polyatomic anions, the rules do not change. Simply treat the polyatomic ion as if it were a single anion. If the cation is from Category I, follow the rules for Category I. If the cation is from Category II, follow the rules for Category II. The crisscross method does not change, either.

EXAMPLE

Use the crisscross method to write the formula for iron(III) sulfate, a Category II cation and a polyatomic ion.

Iron(III) sulfate

$$Fe^{3+} \qquad (SO_4)^{2-}$$
$$Fe_2 \qquad (SO_4)_3$$

The final formula is $Fe_2(SO_4)_3$.

Table 4.7 shows the formulas for some ionic compounds containing common Category I or Category II cations combined with a few common polyatomic ions.

TABLE 4.7 Examples of Ionic Compounds with Polyatomic Ions and Either Category I or II Cations

Name	Formula	Comment
Sodium sulfate	Na_2SO_4	Category I—the Na^+ always is $1+$
Potassium dihydrogen phosphate	KH_2PO_4	The $H_2PO_4^-$ ion has a $1-$ charge and the K^+, from Category I, is $1+$
Iron(III) nitrate	$Fe(NO_3)_3$	Category II—transition metal, must contain a Roman numeral
Cesium perchlorate	$CsClO_4$	The *per*- prefix is used because the polyatomic ion has 1 more oxygen than the chlorate ion
Manganese(II) hydrox-ide	$Mn(OH)_2$	Category II—transition metal, must contain a Roman numeral

Category III—Binary Covalent Compounds

Binary covalent compounds are formed between two nonmetals. Although these compounds do not contain ions, they are named very similarly to binary ionic compounds. To name binary covalent compounds, use these steps.

1. The first element in the formula is named first, using the full elemental name.

2. The second element is named as if it were an anion and uses its elemental name.

3. Prefixes are used to denote the number of the first element unless only one atom of that element is present. For example, CO is called carbon monoxide, not monocarbon monoxide. Many of the prefixes used are shown in Table 4.8 below.

4. Prefixes are always used to denote the number of the second element present.

TABLE 4.8 Prefixes Used to Indicate Numbers in Covalent Compounds

Prefix	Number	Prefix	Number
mono-	1	hexa-	6
di-	2	hepta-	7
tri-	3	octa-	8
tetra-	4	nona-	9
penta-	5	deca-	10

The following are examples of covalent compounds formed from the nonmetals nitrogen and oxygen, using the rules above.

Compound	Systematic Name	Common Name
N_2O	Dinitrogen monoxide*	Nitrous oxide
NO	Nitrogen monoxide*	Nitric oxide
NO_2	Nitrogen dioxide	
N_2O_3	Dinitrogen trioxide	
N_2O_4	Dinitrogen tetroxide*	
N_2O_5	Dinitrogen pentoxide*	

*Notice that for ease of pronunciation, the final "a" or "o" of the prefix is dropped if the element begins with a vowel.

To write the formula for binary covalent compounds, use the same steps as when writing the formula of ionic compounds.

1. The symbol of the first element in the formula is written first, followed by the second element.

2. Use the prefix(es) denoted in the name for the number of each element present in the formula.

The following show some examples of binary covalent compounds.

Name	Formula
Sulfur hexafluoride	SF_6
Phosphorus trichloride	PCl_3

Names and Formulas of Common Acids and Bases

The definition of an acid and a base is expanded later in a first-year chemistry course. For now, common acids are aqueous solutions of hydrogen compounds that contain hydrogen ions, H^+. Common bases are aqueous solutions containing hydroxide ions, OH^-.

A binary acid is named by placing the prefix *hydro-* in front of the stem or full name of the nonmetallic element, and adding the ending *-ic*. Examples are *hydro*chlor*ic* acid (HCl) and *hydro*sulfur*ic* acid (H_2S).

A **ternary compound** consists of three elements, usually an element and a polyatomic ion. To name the compound, you merely name each component in the order of positive first and negative second.

Ternary acids usually contain hydrogen, a nonmetal, and oxygen. Because the amount of oxygen often varies, the name of the most common form of the acid in the series consists of merely the stem of the nonmetal with the ending *-ic*. The acid containing one less atom of oxygen than the most common acid is designated by the ending *-ous*. The name of the acid containing one more atom of oxygen than the most common acid has the prefix *per-* and the ending *-ic*; that of the acid containing one less atom of oxygen than the *-ous* acid has the prefix *hypo-* and the ending *-ous*. This is evident in Table 4.10 with the acids containing H, Cl, and O.

You can remember the names of the common acids and their salts by learning the following simple rules shown in Table 4.9.

TABLE 4.9 Rules for Naming Acids

TIP

Learn these rules.

Rule	Example
-ic acids form *-ate* salts.	Sulfuric acid forms sulfate salts.
-ous acids form *-ite* salts.	Sulfurous acid forms sulfite salts.
hydro-(stem)*-ic* acids form *-ide* salts.	Hydrochloric acid forms chloride salts.

When the name of the ternary acid has the prefix *hypo-* or *per-*, that prefix is retained in the name of the salt (hypochlorous acid = sodium hypochlorite).

The names and formulas of some common acids and bases are listed in Table 4.10.

TABLE 4.10 Formulas of Common Acids and Bases

ACIDS, BINARY		ACIDS, TERNARY	
Name	Formula	Name	Formula
Hydrofluoric	HF	Nitric	HNO_3
Hydrochloric	HCl	Nitrous	HNO_2
Hydrobromic	HBr	Hypochlorous	HClO
Hydriodic	HI	Chlorous	$HClO_2$
Hydrosulfuric	H_2S	Chloric	$HClO_3$
		Perchloric	$HClO_4$
BASES		Sulfuric	H_2SO_4
Sodium hydroxide	NaOH	Sulfurous	H_2SO_3
Potassium "	KOH	Phosphoric	H_3PO_4
Ammonium "	NH_4OH	Phosphorous	H_3PO_3
Calcium "	$Ca(OH)_2$	Carbonic	H_2CO_3
Magnesium "	$Mg(OH)_2$	Acetic	$HC_2H_3O_2$
Barium "	$Ba(OH)_2$	Oxalic	$H_2C_2O_4$
Aluminum "	$Al(OH)_3$	Boric	H_3BO_3
Iron(II) "	$Fe(OH)_2$		
Iron(III) "	$Fe(OH)_3$		
Zinc "	$Zn(OH)_2$		
Lithium "	LiOH		

Chemical Formulas: Their Meaning and Use

As you have seen, a chemical formula is an indication of the makeup of a compound in terms of the kinds of atoms and their relative numbers. It also has some quantitative applications. By using the atomic masses assigned to the elements, we can find the **formula mass** of a compound. If we are sure that the formula represents the actual makeup of one molecule of the substance, the term **molecular mass** may be used as well. In some cases, the formula represents an ionic lattice and no discrete molecule exists, as in the case of table salt, NaCl, or the formula merely represents the simplest ratio of the combined substances and not a molecule of the substance. For example, CH_2 is the simplest ratio of carbon and hydrogen united to form the actual compound ethylene, C_2H_4. This simplest ratio formula is called the **empirical formula**, and the actual formula is the **true formula**. The formula mass is determined by multiplying the atomic mass units (as a whole number) by the subscript for that element and then adding these values for all the elements in the formula. For example:

$Ca(OH)_2$ (one calcium amu + two hydrogen and two oxygen amu = formula mass).

$$1Ca \ (amu = 40) = 40$$
$$2O \ (amu = 16) = 32$$
$$2H \ (amu = 1) = 2.0$$
$$\overline{\text{Formula mass } Ca(OH)_2 = 74 \text{ amu (or } \mu)}$$

It is sometimes useful to know what percent of the total weight of a compound is made up of a particular element. This is called finding the **percentage composition**. The simple formula for this is:

$$\frac{\text{Total amu of the element in the compound}}{\text{Total formula amu}} \times 100\% = \text{Percentage composition of that element}$$

To find the percent composition of calcium in calcium hydroxide in the example above, we set the formula up as follows:

$$\frac{Ca = 40.amu}{\text{Formula mass} = 74 \text{ amu}} \times 100\% = 54\% \text{ Calcium}$$

To find the percent composition of oxygen in calcium hydroxide:

$$\frac{O = 32 \text{ amu}}{\text{Formula mass} = 74 \text{ amu}} \times 100\% = 43\% \text{ Oxygen}$$

To find the percent composition of hydrogen in calcium hydroxide:

$$\frac{H = 2.0 \text{ amu}}{\text{Formula mass} = 74 \text{ amu}} \times 100\% = 2.7\% \text{ Hydrogen}$$

EXAMPLE

Given: Ba = 58.81%, S = 13.73%, and O = 27.46%.

Find the empirical formula.

1. Divide each percent by the amu of the element.

Ba	S	O
$\frac{58.8}{137} = 0.43$	$\frac{13.7}{32} = 0.43$	$\frac{27.5}{16} = 1.72$

2. Manipulate numbers to get small whole numbers. Try dividing them all by the smallest first. In this case, divide each result by 0.43, as shown below.

Ba	S	O
$\frac{0.43}{0.43} = 1$	$\frac{0.43}{0.43} = 1$	$\frac{1.72}{0.43} = 4$

3. The formula is $BaSO_4$.

In some cases, you may be given the true formula mass of the compound. To check if your empirical formula is correct, add up the formula mass of the empirical formula and compare it with the given formula mass. If it is *not* the same, multiply the empirical formula by the small whole number that gives you the correct formula mass. For example, if your empirical formula is CH_2 (which has a formula mass of 14) and the true formula mass is given as 28, you can see that you must double the empirical formula by doubling all the subscripts. The true formula is C_2H_4.

Laws of Definite Composition and Multiple Proportions

In the problems involving percent composition, we have depended on two things: each unit of an element has the same atomic mass, and every time the particular compound forms, it forms in the same percent composition. That this latter statement is true no matter the source of the compound is the **Law of Definite Composition**. There are some compounds formed by the same two elements in which the mass of one element is constant, but the mass of the other varies. In every case, however, the mass of the other element is present in a small-whole-number ratio to the weight of the first element. This is called the **Law of Multiple Proportions**. An example is H_2O and H_2O_2.

In H_2O the proportion of H : O = 2 : 16 or 1 : 8

In H_2O_2 the proportion of H : O = 2 : 32 or 1 : 16

The ratio of the mass of oxygen in each is 8 : 16 or 1 : 2 (a small-whole-number ratio).

Writing and Balancing Simple Equations

An equation is a simplified way of recording a chemical change. Instead of words, chemical symbols and formulas are used to represent the **reactants** and the **products**. Here is an example of how this can be done. The following is the word equation of the reaction of burning hydrogen with oxygen:

Hydrogen + oxygen yields water

Replacing the words with the chemical formulas, we have:

$$H_2 + O_2 \rightarrow H_2O$$

We replaced hydrogen and oxygen with the formulas for their diatomic molecular states and wrote the appropriate formula for water based on the respective oxidation (valence) numbers for hydrogen and oxygen. Note that the word **yields** was replaced with the arrow.

Although the chemical statement tells what happened, it is not an equation because the two sides are not equal. While the left side has two atoms of oxygen, the right side has only one. Knowing that the Law of Conservation of Matter dictates that matter cannot easily be created or destroyed, we must get the number of atoms of each element represented on the left side to equal the number on the right. To do this, we can only use numbers, called **coefficients**, in front of the formulas. It is important to note that in attempting to balance equations, THE SUBSCRIPTS IN THE FORMULAS MAY NOT BE CHANGED.

Looking again at the skeleton equation, we notice that if 2 is placed in front of H_2O, the numbers of oxygen atoms represented on the two sides of the equation are equal. However, there are now four hydrogens on the right side with only two on the left. This can be corrected by using a coefficient of 2 in front of H_2. Now we have a balanced equation:

$$2H_2 + O_2 \rightarrow 2H_2O$$

This equation tells us more than merely that hydrogen reacts with oxygen to form water. It has quantitative meaning as well. It tells us that two molecular masses of hydrogen react with one molecular mass of oxygen to form two molecular masses of water. Because molecular masses are indirectly related to grams, we may also relate the masses of reactants and products in grams.

$2H_2$	+	O_2	$\rightarrow$	$2H_2O$
2(2)		32		2(18)
4 units	+	32 units	=	36 units
4 grams of H_2	+	32 grams of O_2	=	36 grams of water

This aspect will be important in solving problems related to the masses of substances in a chemical equation.

TIP
You cannot change subscripts of formulas to attempt to balance an equation!

Showing Phases in Chemical Equations

Once an equation is balanced, you may choose to give additional information in the equation. This can be done by indicating the phases of substances, telling whether each substance is in the liquid phase (ℓ), the gaseous phase (g), or the solid phase (s). Since many solids will not react to any appreciable extent unless they are dissolved in water, the notation (aq) is used to indicate that the substance exists in a water (aqueous) solution. Information concerning phase is given in parentheses following the formula for each substance. Several illustrations of this notation are given in Table 4.11 below.

TABLE 4.11 Examples of Phase Notation

Formula with Phase Notation	Meaning
$Cl_2(g)$	Chlorine gas
$H_2O(\ell)$	Water in the liquid state as opposed to ice or steam
NaCl(s)	Sodium chloride as a solid
NaCl(aq)	A water solution of dissolved sodium chloride

TIP

(g) = gaseous state

(ℓ) = liquid state

(s) = solid state

An example of phase notation in an equation:

$$2HCl(aq) + Zn(s) \rightarrow ZnCl_2(aq) + H_2(g)$$

In words, this says that a water solution of hydrogen chloride (called hydrochloric acid) reacts with solid zinc to produce zinc chloride dissolved in water plus hydrogen gas.

Writing Ionic Equations

At times, chemists choose to show only the substances that react in the chemical action. These equations are called **ionic** equations because they stress the reaction and production of ions. If we look at the preceding equation, we see the complete cast of "actors":

TIP

In net ionic equations, do not show "spectator" ions that do not change.

Reactants

2HCl(aq) releases $\rightarrow$ $2H^+$ (aq) + $2Cl^-$ (aq) in solution

Zn(s) stay as $\rightarrow$ Zn(s) particles

Products

$ZnCl_2$ (aq) releases $\rightarrow$ Zn_2 + (aq) + $2Cl^-$ (aq) in solution

H_2 (g) stay as $\rightarrow$ H_2 (g) molecules

Writing the complete reaction using these results, we have:

$$2H^+(aq) + 2Cl^-(aq) + Zn(s) \rightarrow Zn^{2+}(aq) + 2Cl^-(aq) + H_2(g)$$

Notice that nothing happened to the chloride ion. It appears the same on both sides of the equation. It is referred to as a spectator ion. In writing the **net ionic equation**, spectator ions are omitted, so the net ionic equation is:

$$2H^+(aq) + Zn(s) \rightarrow Zn^{2+}(aq) + H_2(g)$$

CHAPTER 5

Gases and the Gas Laws

> ## Learning Objectives
>
> In this chapter, you will learn how to:
>
> - Explain how atmospheric pressure is measured, how to read the pressure in a manometer, and the units used to measure pressure.
> - Read and explain a graphic distribution of the number of molecules versus kinetic energy at different temperatures.
> - Know and use the following laws to solve gas problems: Graham's, Charles's, Boyle's, Dalton's, the Combined Gas Law, and the Ideal Gas Law.

When we discuss gases today, the most pressing concern is the gases in our atmosphere. These are the gases that are held against Earth by the gravitational field. The principal constituents of the atmosphere of Earth today are nitrogen (78%) and oxygen (21%). The gases in the remaining 1% are argon (0.9%), carbon dioxide (0.03%), varying amounts of water vapor, and trace amounts of hydrogen, ozone, methane, carbon monoxide, helium, neon, krypton, and xenon.

Studies of air samples show that up to 55 miles above sea level, the composition of the atmosphere is substantially the same as at ground level; continuous stirring produced by atmospheric currents counteracts the tendency of the heavier gases to settle below the lighter ones. In the lower atmosphere, ozone is normally present in extremely low concentrations. The atmospheric layer 12 to 30 miles up contains more ozone that is produced by the action of ultraviolet radiation from the Sun. In this layer, however, the percentage of ozone is only 0.001 by volume. Human activity adds to the ozone concentration in the lower atmosphere where it can be a harmful pollutant.

> **TIP**
> The major components of Earth's atmosphere: 78% nitrogen 21% oxygen
>
>

General Characteristics of Gases

The state of any substance is defined by various measurable parameters associated with that substance. These include pressure, volume, temperature, and the number of particles present in a sample of the substance. Knowledge of these parameters for gases allows chemists to predict the manner in which the gaseous substances will behave.

Measuring the Pressure of a Gas

Pressure is defined as force per unit area. With respect to the atmosphere, pressure is the result of the weight of a mixture of gases. This pressure, which is called **atmospheric pressure**, **air pressure**, or **barometric pressure**, is approximately equal to the weight of a kilogram mass on every square centimeter of surface exposed to it. This weight is about 10 newtons.

The pressure of the atmosphere varies with altitude. At higher altitudes, the weight of the overlying atmosphere is less, so the pressure is less. Air pressure also varies somewhat with weather conditions as low- and high-pressure areas move with weather fronts. On the average, however, the air pressure at sea level can support a column of mercury 760 millimeters in height. This average sea-level air pressure is known as **normal atmospheric pressure**, also called **standard pressure**.

The instrument most commonly used for measuring air pressure is the **mercury barometer**. Figure 5.1 below shows how it operates. Atmospheric pressure is exerted on the mercury in the dish, and this in turn holds the column of mercury up in the tube. This column at standard pressure will measure 760 millimeters above the level of the mercury in the dish below.

TIP

Read the top of the meniscus for mercury but the bottom of the meniscus for water.

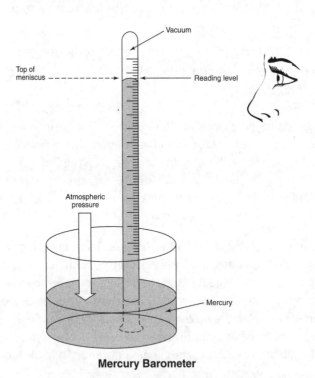

Mercury Barometer

FIGURE 5.1 A Barometer Reading

In gas-law problems, pressure may be expressed in various units. One standard atmosphere (1 atm) is equal to 760 millimeters of mercury (760 mm Hg) or 760 **torr**, a unit named for Evangelista Torricelli. In the SI system, the unit of pressure is the **pascal** (Pa), named in honor of the scientist of the same name, and standard pressure is 101,325 pascals or 101.325 kilopascals (kPa). One pascal (Pa) is defined as the pressure exerted by the force of one newton (1 N) acting on an area of one square meter. In many cases, as in atmospheric pressure, it is more convenient to express pressure in kilopascals (kPa). Table 5.1 lists these typical units of pressure and the quantity of each that is equal to 1 atmosphere.

TABLE 5.1 Summary of Units of Pressure

Unit	Abbreviation	Unit Equivalent to 1 atm
Atmosphere	atm	1 atm
Millimeters of Hg	mm Hg	760 mm Hg
Torr	torr	760 torr
Pascal	Pa	101,325 Pa
Kilopascal	kPa	101.325 kPa

A device similar to the barometer can be used to measure the pressure of a gas in a confined container. This apparatus, called a **manometer**, is illustrated below in Figure 5.2. A manometer is basically a U-tube containing mercury or some other liquid. When both ends are open to the air, as in (1) in the diagram, the level of the liquid will be the same on both sides since the same pressure is being exerted on both ends of the tube. In (2) and (3), a vessel is connected to one end of the U-tube. Now the height of the mercury column serves as a means of reading the pressure inside the vessel if the atmospheric pressure is known. When the pressure inside the vessel is the same as the atmospheric pressure outside, the levels of liquid are the same. When the pressure inside is greater than outside, the column of liquid will be higher on the side that is exposed to the air, as in (2). Conversely, when the pressure inside the vessel is less than the outside atmospheric pressure, the additional pressure will force the liquid to a higher level on the side near the vessel, as in (3).

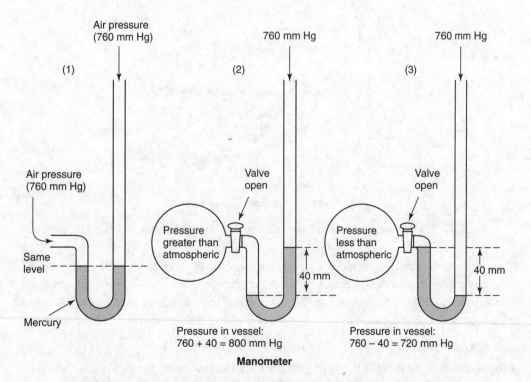

FIGURE 5.2 Three Scenarios for Measurements with a Manometer

Kinetic-Molecular Theory

By indirect observations, the **Kinetic-Molecular Theory** has been arrived at to explain the forces between molecules and the energy the molecules possess. There are three basic assumptions to the Kinetic-Molecular Theory:

1. Matter in all its forms (solid, liquid, and gas) is composed of extremely small particles. In many cases these are called molecules. The space occupied by the gas particles themselves is ignored in comparison with the volume of the space in which they are contained.

2. The particles of matter are in constant motion. In solids, this motion is restricted to a small space. In liquids, the particles have a more random pattern but still are restricted to a kind of rolling over one another. In a gas, the particles are in continuous, random, straight-line motion.

3. When these particles collide with each other or with the walls of the container, there is no loss of energy.

Some Particular Properties of Gases

As the temperature of a gas is increased, its kinetic energy is increased, thereby increasing the random motion. At a particular temperature, not all the particles have the same kinetic energy, but the temperature is a measure of the average kinetic energy of the particles. A graph of the various kinetic energies resembles a normal bell-shaped curve with the average found at the peak of the curve (see Figure 5.3).

TIP

When you read the temperature of a substance, you are measuring its average kinetic energy.

TIP

Diffusion means spreading out.

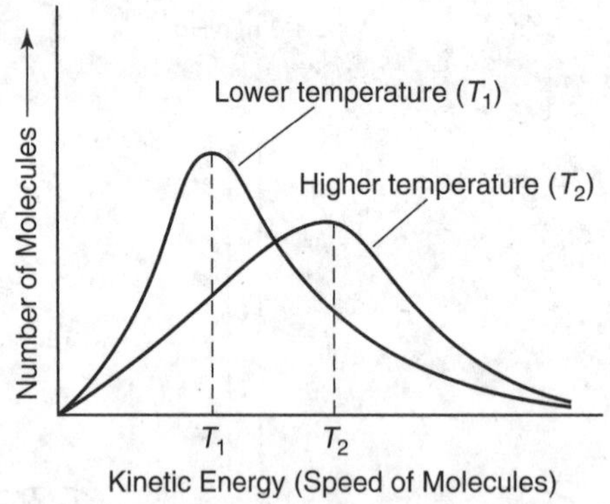

FIGURE 5.3 Molecular Speed Distribution in a Gas at Different Temperatures

When the temperature is lowered, the gas reaches a point at which the kinetic energy can no longer overcome the attractive forces between the particles (or molecules) and the gas condenses to a liquid. The temperature at which this condensation occurs is related to the type of substance the gas is composed of and the type of bonding in the molecules themselves.

The random motion of gases in moving from one position to another is referred to as **diffusion**. You know that if a bottle of perfume is opened in one corner of a room, the perfume, that is, its molecules, will move or diffuse to all parts of the room in time. The rate of diffusion is the rate of the mixing of gases.

Effusion is the term used to describe the passage of a gas through a tiny orifice into an evacuated chamber. The rate of effusion measures the speed at which the gas is transferred into the chamber.

Gas Laws and Related Problems

The manner in which gases behave is predictable and is based on observations that have been seen over and over again. These are called laws. Many have names based on the scientist responsible for first recognizing and articulating the behavior.

Graham's Law of Effusion (Diffusion)

This law relates the rate at which a gas diffuses (or effuses) to the type of molecule in the gas. It can be expressed as follows:

> The rate of effusion of a gas is inversely proportional to the square root of its molecular mass.

Hydrogen, with the lowest molecular mass, can diffuse more rapidly than other gases under similar conditions.

EXAMPLE

Compare the rate of diffusion of hydrogen to that of oxygen under similar conditions.

The formula is:

$$\frac{\text{Rate A}}{\text{Rate A}} = \frac{\sqrt{\text{Molecular mass of B}}}{\sqrt{\text{Molecular mass of A}}}$$

Let A be H_2 and B be O_2.

$$\frac{\text{Rate } H_2}{\text{Rate } O_2} = \frac{\sqrt{32}}{\sqrt{2}} = \frac{\sqrt{16}}{\sqrt{1}} = \frac{4}{1}$$

Therefore hydrogen diffuses four times as fast as oxygen.

In dealing with the gas laws, a student must know what is meant by standard conditions of temperature and pressure (abbreviated as STP). The standard pressure is defined as the height of mercury that can be held in an evacuated tube by 1 atmosphere of pressure (14.7 lb./in.^{2}). This is usually expressed as 760 millimeters of Hg or 101.3 kilopascals. **Standard temperature** is defined as 273 Kelvin or absolute (which corresponds to 0° Celsius).

Charles's Law $\left(\frac{V}{T} = k\right)$

Jacques Charles, a French chemist of the early nineteenth century, discovered that, when a gas under constant pressure is heated from 0°C to 1°C, it expands 1/273 of its volume. It contracts this amount when the temperature is dropped 1 degree to −1°C. Charles reasoned that, if a gas at 0°C was cooled to −273°C (actually found to be −273.15°C), its volume would be zero. Actually, all gases are converted into liquids before this temperature is reached. By using the Kelvin scale to rid the problem of negative numbers, we can state **Charles's Law** as follows:

> If the pressure remains constant, the volume of a gas varies directly as the absolute temperature. Then:
>
> $$\text{Initial } \frac{V_1}{T_1} = \text{Final } \frac{V_2}{T_2} \text{ at constant pressure or } \frac{V}{T} = k$$

TIP

Charles's Law $\frac{V_1}{T_1} = \frac{V_2}{T_2}$ at constant pressure is a direct proportion.

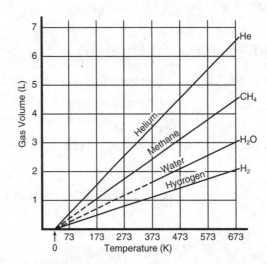

FIGURE 5.4 Plots of V vs. T for Representative Gases

TIP

Assume dry gases unless otherwise stated.

A graphical relationship of Charles's Law is shown above in Figure 5.4. The dashed lines represent extrapolation of the data into regions where the gas would become liquid or solid. Extrapolation shows that each gas, if it remained gaseous, would reach zero volume at 0 K or −273°C.

EXAMPLE

The volume of a gas at 20°C is 500 mL. Find its volume at standard temperature if pressure is held constant.

Convert temperatures:

$$20°C = 20° + 273 = 293 \text{ K}$$
$$0°C = 0° + 273 = 273 \text{ K}$$

If you know that cooling a gas decreases its volume, then you know that 500 mL will have to be multiplied by a fraction (made up of the Kelvin temperatures) that has a smaller numerator than the denominator. So:

$$500 \text{ mL} \times \frac{273}{293} = 465 \text{ mL}$$

Or you can use the formula and substitute known values:

$$\frac{V_1}{T_1} = \frac{V_2}{T_2}$$
$$\frac{500 \text{ mL}}{293} = \frac{x \text{ mL}}{273}$$
$$x \text{ mL} = 465 \text{ mL}$$

TIP

STP = standard temperature of 273 K standard pressure of 760 mm Hg or 1 atmosphere (atm) or 101.3 kilopascals.

Boyle's Law

Robert Boyle, a seventeenth-century English scientist, found that the volume of a gas decreases when the pressure on it is increased, and vice versa, when the temperature is held constant. **Boyle's Law** can be stated as follows:

If the temperature remains constant, the volume of a gas varies inversely as the pressure changes. Then:

$$P_1V_1 = P_2V_2 \text{ at a constant temperature}$$

or

$$PV = k$$

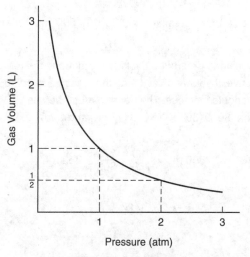

Relationship—Boyle's Law

FIGURE 5.5 A Plot of V vs. P for a Typical Gas at a Constant Temperature

A graphical relationship of Boyle's Law is shown above in Figure 5.5. The curve is referred to as a hyperbola and describes an inverse relationship between pressure and volume.

EXAMPLE

Given the volume of a gas as 200. mL at 1.05 atm pressure, calculate the volume of the same gas at 1.01 atm. Temperature is held constant.

If you know that this decrease in pressure will cause an increase in the volume, then you know 200 mL must be multiplied by a fraction (made up of the two pressures) that has a larger numerator than the denominator. So:

$$200 \text{ mL} \times \frac{1.05 \text{ atm}}{1.01 \text{ atm}} = 208 \text{ mL}$$

Or you can use the formula:

$$P_1V_1 = P_2V_2$$

$$V_2 = V_1 \times \frac{P_1}{P_2}$$

$$= 200 \text{ mL} \times \frac{1.05 \text{ atm}}{1.01 \text{ atm}} = 208 \text{ mL}$$

Combined Gas Law

The **combined gas law** is a combination of the two preceding gas laws. The formula is as follows:

$$\frac{P_1V_1}{T_1} = \frac{P_2V_2}{T_2}$$

EXAMPLE

The volume of a gas at 780 mm pressure and 30°C is 500 mL. What volume would the gas occupy at STP?

You again can use reasoning to determine the kind of fractions the temperatures and pressures must be to arrive at your answer. Since the pressure is going from 780 mm to 760 mm, the volume should increase. The fraction must then be $\frac{780}{760}$. Also, since the temperature is going from 30°C (303 K) to 0°C (273 K), the volume should decrease; this fraction must be $\frac{273}{303}$. So:

$$500 \text{ mL} \times \frac{780}{760} \times \frac{273}{303} = 462 \text{ mL}$$

Or you can use the formula:

$$\frac{P_1V_1}{T_1} = \frac{P_2V_2}{T_2}$$

Solve for $V_2 = V_1 \times \frac{P_1}{P_2} \times \frac{T_2}{T_1}$

$$V_2 = 500 \text{ mL} \times \frac{780 \text{ mm Hg}}{760 \text{ mm Hg}} \times \frac{273 \text{ K}}{303 \text{ K}} = 462 \text{ mL}$$

Pressure versus Temperature (Gay-Lussac's Law)

At constant volume, the pressure of a given mass of gas varies directly with the absolute temperature. Then:

$$\frac{P_1}{T_1} = \frac{P_2}{T_2} \text{ at constant volume or } \frac{P_1}{T_1} = k$$

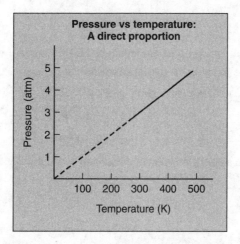

FIGURE 5.6 A Plot of *P* vs. *T* for a Typical Gas at Constant Volume

A graphical relationship of **Gay-Lussac's Law** is shown above in Figure 5.6. The dashed lines represent extrapolation of the data into regions where the gas would become liquid or solid. Extrapolation shows that each gas, if it remained gaseous, would reach zero volume at 0 K or −273°C.

<hr>

EXAMPLE

A steel tank contains a gas at 27°C and a pressure of 12.0 atms. Determine the gas pressure when the tank is heated to 100°C.

Reasoning that an increase in temperature will cause an increase in pressure at constant volume, you know the pressure must be multiplied by a fraction that has a larger numerator than denominator. The fraction must be $\frac{373\ K}{300\ K}$. So

$$12.0\ \text{atm} \times \frac{373\ \cancel{K}}{300\ \cancel{K}} = 14.9\ \text{atm or } 15.0\ \text{atm}$$

Or you can use the formula:

$$\frac{P_1}{T_1} = \frac{P_2}{T_2}$$

$$P_2 = P_1 \times \frac{T_2}{T_1}$$

$$P_2 = 12.0\ \text{atm} \times \frac{373\ K}{300\ K} = 14.9\ \text{atm}$$

Dalton's Law of Partial Pressures

When a gas is made up of a mixture of different gases, the total pressure of the mixture is equal to the sum of the partial pressures of the components; that is, the partial pressure of the gas would be the pressure of the individual gas if it alone occupied the volume. The formula is:

$$P_{\text{total}} = P_{\text{gas 1}} + P_{\text{gas 2}} + P_{\text{gas 3}} + \cdots$$

A mixture of gases at 760 mm Hg pressure contains 65.0% nitrogen, 15.0% oxygen, and 20.0% carbon dioxide by volume. What is the partial pressure of each gas?

$$0.650 \times 760 = 494 \text{ mm pressure } (N_2)$$
$$0.150 \times 760 = 114 \text{ mm pressure } (O_2)$$
$$0.200 \times 760 = 152 \text{ mm pressure } (CO_2)$$

If the pressure was given as 1.0 atm, you would substitute 1.0 atm for 760 mm Hg. The answers would be:

$$0.650 \times 1.0 \text{ atm} = 0.650 \text{ atm } (N_2)$$
$$0.150 \times 1.0 \text{ atm} = 0.150 \text{ atm } (O_2)$$
$$0.200 \times 1.0 \text{ atm} = 0.200 \text{ atm } (CO_2)$$

Corrections of Pressure

Correction of Pressure When a Gas Is Collected Over Water

When a gas is collected over a volatile liquid, such as water, some of the water vapor is present in the gas and contributes to the total pressure. Assuming that the gas is saturated with water vapor at the given temperature, you can find the partial pressure due to the water vapor in a table of such water vapor values. This vapor pressure, which depends only on the temperature, must be subtracted from the total pressure to find the partial pressure of the gas being measured.

Correction of Difference in the Height of the Fluid

When gases are collected in eudiometers (glass tubes closed at one end), it is not always possible to get the level of the liquid inside the tube to equal the level on the outside. This deviation of levels must be taken into account when determining the pressure of the enclosed gas. There are then two possibilities: (1) When the level inside is higher than the level outside the tube, the pressure on the inside is less, by the height of fluid in excess, than the outside pressure. If the fluid is mercury, you simply subtract the difference from the outside pressure reading (also in height of mercury and in the same units) to get the corrected pressure of the gas. If the fluid is water, you must first convert the difference to an equivalent height of mercury by dividing the difference by 13.6 (since mercury is 13.6 times as heavy as water, the height expressed in terms of Hg will be 1/13.6 the height of water). This is shown in Figure 5.7. Again, care must be taken that this equivalent height of mercury is in the same units as the expression for the outside pressure before it is subtracted to obtain the corrected pressure for the gas in the eudiometer. (2) When the level inside is lower than the level outside the tube, a correction must be added to the outside pressure. If the difference in height between the inside and the outside is expressed in terms of water, you must take 1/13.6 of this quantity to correct it to millimeters of mercury. This quantity is then added to the expression of the outside pressure, which must also be in millimeters of mercury. If the tube contains mercury, then the difference between the inside and outside levels is merely added to the outside pressure to get the corrected pressure for the enclosed gas.

TIP

When a gas is collected over water, subtract the water vapor pressure at the given temperature from the atmospheric pressure to find the partial pressure of the gas. $P_{gas} = P_{atm} - P_{H_2O}$

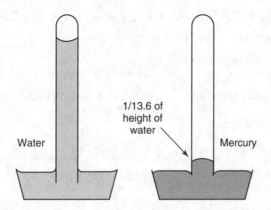

FIGURE 5.7 Same Pressure Exerted on Both Liquids

> ### EXAMPLE
>
> Hydrogen gas was collected in a eudiometer tube over water. It was impossible to level the outside water with that in the tube, so the water level inside the tube was 40.8 mm higher than that outside. The barometric pressure was 730 mm of Hg. The water vapor pressure at the room temperature of 29°C was found in a handbook to be 30.0 mm. What is the pressure of the dry hydrogen?
>
> **STEP 1** To find the true pressure of the gas, we must first subtract the water-level difference expressed in mm of Hg:
>
> $$\frac{40.8}{13.6} = 3.00 \text{ mm of Hg}$$
>
> Then 730 mm − 3.00 mm = 727 mm total pressure of gases in the eudiometer.
>
> **STEP 2** Correcting for the partial pressure due to water vapor in the hydrogen, we subtract the vapor pressure (30.0 mm) from 727 mm and get our answer: 697 mm.

Ideal Gas Law

The preceding laws do not include the relationship of number of moles of a gas to the pressure, volume, and temperature of the gas. A law derived from the Kinetic-Molecular Theory relates these variables. It is called the **Ideal Gas Law** and is expressed as:

$$PV = nRT$$

P, V, and T retain their usual meanings, but n stands for the number of moles of the gas and R represents the ideal gas constant.

Boyle's Law and Charles's Law are actually derived from the Ideal Gas Law. Boyle's Law applies when the number of moles and the temperature of the gas are constant. Then in $PV = nRT$, the number of moles, n, is constant; the gas constant (R) remains the same; and by definition T is constant. Therefore, $PV = k$. At the initial set of conditions of a problem, $P_1V_1 = $ a constant (k). At the second set of conditions, the terms on the right side of the equation are equal to the same constant, so $P_1V_1 = P_2V_2$. This matches the Boyle's Law equation introduced earlier.

The same can also be done with Charles's Law, because $PV = nRT$ can be expressed with the variables on the left and the constants on the right:

$$\frac{V}{T} = \frac{nR}{P}$$

In Charles's Law, the number of moles and the pressure are constant. Substituting k for the constant term, $\frac{nR}{P}$, we have:

$$\frac{V}{T} = k$$

The expression relating two sets of conditions can be written as:

$$\frac{V_1}{T_1} = \frac{V_2}{T_2}$$

To use the ideal gas law in the form $PV = nRT$, the gas constant, R, must be determined. This can be done mathematically as shown in the following example.

TIP

Remember to use appropriate units:

moles (mol)

liters (L)

atmosphere (atm)

EXAMPLE

One mole of oxygen gas was collected in the laboratory at a temperature of 24.0°C and a pressure of exactly 1 atmosphere. The volume was 24.38 liters. Find the value of R.

$$PV = nRT$$

Rearranging the equation to solve for R gives:

$$R = \frac{PV}{nT}$$

Substituting the known values on the right, we have:

$$R = \frac{1\,\text{atm} \times 24.38\ \text{L}}{1\ \text{mol} \times 297\ \text{K}}$$

Calculating R, we get:

$$R = 0.0821\ \frac{\text{L} \cdot \text{atm}}{\text{mol} \cdot \text{K}}$$

Once R is known, the Ideal Gas Law can be used to find any of the variables, given the other three.

CHAPTER 6

Stoichiometry (Chemical Calculations) and the Mole Concept

This chapter deals with the solving of a variety of quantitative chemistry problems, which is often referred to as **stoichiometry**.

Although there are many specific methods for solving the types of problems in this chapter, we will focus primarily on the technique called **dimensional analysis**. Dimensional analysis provides a very clear problem-solving pathway for stoichiometry problems. It emphasizes not only the numerical values involved in the calculations but also the units describing the quantities in question.

The Mole Concept

Providing a name for a quantity of things taken as a whole is common in everyday life. Some examples are a dozen, a gross, and a ream. Each of these represents a specific number of items and is not dependent on the commodity. A dozen eggs, oranges, or bananas will always represent 12 items. In chemistry, we have a unit that decribes a quantity of particles. It is called the **mole** (sometimes abbreviated as **mol**). A mole is 6.02×10^{23} particles. Technically, that's the number of carbon atoms found in exactly 12 grams of carbon-12. Since the atomic masses of all the elements' atoms are related to the mass of carbon-12, a mole is also the number of atoms found in the atomic mass of *any* element if it is *expressed in grams*. Keep in mind that the masses found on the Periodic Table for any element are actually weighted averages of all the isotopes that exist for that element (based on their relative natural abundances). Those masses, if expressed in atomic mass units (amu), represent just one average atom for that element. If, however, the value for mass was expressed in grams, that sample of the element would contain 6.02×10^{23} atoms of that element. This value is also known as **Avogadro's number** in honor of the Italian scientist whose hypothesis concerning the volumes of gases led to its determination. More on Avogadro's hypothesis will be discussed in an upcoming section on gas volumes and molar mass. Avogadro's number is very large because the items being counted (atoms) are very small. So 6.02×10^{23} atoms of most elements represent samples of atoms that are conveniently sized for working in the laboratory.

Molar Mass and Moles

The mass of a mole of particles is referred to as its **molar mass**. For moles of atoms, the atomic mass found on the Periodic Table for that element expressed in grams is the molar mass for that element. Some elements naturally exist as molecules, however. The molar mass of those elements takes into account the number of atoms in the molecule *in an additive manner*. Most elements are considered in a monatomic way (one atom). However, a few (hydrogen, nitrogen, oxygen, fluorine, chlorine, bromine, and iodine) are typically considered in a diatomic manner (two atoms) based on the way they are generally found to exist. This is not to say that you could not count moles of hydrogen atoms (H) as opposed to hydrogen molecules (H_2). You should, though, always be cognizant of the type of particle involved in any mole calculation.

EXAMPLE

Determine the molar mass of silicon, using the Periodic Table.

The atomic mass of silicon is 28.1 amu as found on the Periodic Table. Therefore, the molar mass of silicon is 28.1 g and represents 6.02×10^{23} *atoms* of silicon or 1 mole of silicon atoms.

Elements are just one type of substance for which the molar mass can be found. The molar masses of compounds can be found in a way similar to that of diatomic elements. Just add up the molar masses of the individual elements found in the compound based on the compound's formula.

EXAMPLE

Find the molar mass of NH_3 (this is a molecular compound known as ammonia).

The molar mass of nitrogen is 14.0 g, and the molar mass of hydrogen is 1.0 g. Since there are 3 hydrogen atoms per molecule of ammonia, the molar mass of ammonia is 17.0 g and represents 6.02×10^{23} molecules of ammonia or 1 mole of ammonia molecules.

Molar Mass and Gas Volumes

In 1811, Amedeo Avogadro made a far-reaching scientific assumption that also bears his name. **Avogadro's Hypothesis** states that equal volumes of different gases contain equal numbers of particles at the same temperature and pressure. It means that under the same conditions, the number of molecules of hydrogen in a 1-liter container is exactly the same as the number of molecules of carbon dioxide (or of any other gas) in a 1-liter container even though the individual molecules of the different gases have different masses and sizes. Because of the substantiation of this hypothesis by much data since its inception, it is often referred to as **Avogadro's Law**. Avogadro's Law shows the relationship between the volume and the number of particles of a gas sample when the temperature and pressure are constant:

$$\frac{V}{n} = k$$

In other words, volume and the number of gas particles are directly related.

Because the volume of a gas may vary depending on the temperature and pressure, a standard is set for comparing gases. The standard conditions of temperature and pressure (abbreviated STP) are 273 K and 1 atmosphere. Because the relationship between volume and number of particles of a gas is direct when the temperature and pressure are constant, the molar mass of a gas (which represents a *set* number of particles, namely 1 mole) occupies a *set* volume. The volume of 22.4 L is recognized as the molar volume of any gas at STP.

Using Molar Mass and Molar Volume

Molar mass and molar volume are typically used as conversion factors to change quantities of reactants expressed as masses or as volumes to moles for use in stoichiometry problems via dimensional analysis.

EXAMPLE

Find the number of moles of silicon present in 4.30 g of silicon.

Recall that silicon is an element that is not diatomic. It has a molar mass of 28.1 g. So 28.1 g of silicon contains 1.00 mol of silicon atoms if significant figures are kept in mind.

Use dimensional analysis:

$$\frac{4.30 \text{ g Si}}{1} \times \frac{1.00 \text{ mol Si}}{28.1 \text{ g Si}} = 0.0153 \text{ mol Si}$$

Note that when using dimensional analysis, the given quantity is simply multiplied by a factor that is equal to the value 1 since the numerator and denominator in that factor are equal to each other. Using this method causes the magnitude of the given quantity not to change. What does change is the unit in which the quantity is expressed. The given units cancel out, leaving only the unit in the numerator of the factor to describe the quantity.

Density and Molar Mass

Since the density of a gas is usually given in grams/liter of gas at STP, we can use the molar volume at STP to solve the following types of problems.

EXAMPLE

Find the molar mass of a gas when the density is given as 1.25 grams/liter.

Because it is known that 1 mole of a gas occupies 22.4 liters at STP, we can solve this problem by multiplying the mass of 1 liter by 22.4 liters/mole using dimensional analysis:

$$\frac{1.25 \text{ g}}{\cancel{L}} \times \frac{22.4 \cancel{L}}{1 \text{ mol}} = 28.0 \text{ g/mol}$$

Even if the mass given is not for 1 liter, the same setup can be used.

Stoichiometry: Mole-Mole Problems

The types of mole problems investigated so far have been ones involving only one substance. Chemical calculations often involve more than one substance and take into consideration information found in **balanced reaction equations**. Recall that coefficients from a balanced equation can be used to describe the numbers of atoms, ions, or molecules involved in the chemical process. Also recall that the quantities of each must balance on both sides of the equation to satisfy the Law of Conservation of Matter. Those coefficients can also represent larger numbers of particles, namely moles of those substances, that are reacting or being produced. Therefore, the balanced chemical reaction

$$2 \text{ NaClO(s)} \rightarrow 2 \text{ NaCl(s)} + O_2 \text{ (g)}$$

can be interpreted in two ways.

1. Decomposing 2 formula units of sodium hypochlorite produces 2 formula units of sodium chloride and 1 molecule of oxygen.
2. Decomposing 2 moles of sodium hypochlorite produces 2 moles of sodium chloride and 1 mole of oxygen molecules.

Coefficients in balanced chemical reaction equations therefore provide **mole ratios** for reacting substances and substances produced.

EXAMPLE

How many moles of sodium chloride can be produced from 0.0253 moles of sodium hypochlorite?

Use dimensional analysis:

$$\frac{0.0253 \text{ mol NaClO}}{1} \times \frac{2 \text{ mol NaCl}}{2 \text{ mol NaClO}} = 0.0253 \text{ mol NaCl}$$

Stoichiometry: Mass-Mass Problems

In order to work with substances in the laboratory, chemists must work with quantities they can easily measure. A mole is an amount impractical to actually count out in the lab. Using molar mass allows problems to be based on mass, which is a measurable quantity. **Mass-mass** problems often involve determining the masses of other substances needed to react with a given mass of a substance or the mass of other substances that can be produced from that given mass.

EXAMPLE

Consider the following balanced chemical reaction equation:

$$CaCO_3 \text{ (s)} + 2 \text{ HCl(aq)} \rightarrow CaCl_2 \text{ (aq)} + H_2O(l) + CO_2 \text{ (g)}$$

If 2.59 g of $CaCO_3$ was reacted with enough HCl to use up all of the $CaCO_3$, what mass of HCl would be needed and what mass of CO_2 would be produced?

Use dimensional analysis:

$$\frac{2.59 \text{ g } CaCO_3}{1} \times \underbrace{\frac{1 \text{ mol } CaCO_3}{100.1 \text{ g } CaCO_3}}_{\uparrow \text{Factor\#1}} \times \underbrace{\frac{2 \text{ mol HCl}}{1 \text{ mol } CaCO_3}}_{\uparrow \text{Factor\#2}} \times \underbrace{\frac{36.5 \text{ g HCl}}{1 \text{ mol HCl}}}_{\uparrow \text{Factor\#3}} = 1.89 \text{ g HCl}$$

The first factor in the dimensional analysis equation converts the mass of the $CaCO_3$ to moles by using information concerning molar mass as found from the Periodic Table. The second factor converts moles of $CaCO_3$ to moles of HCl needed to react with the $CaCO_3$. The mole ratio was found from the coefficients in front of each substance in the balanced reaction. The third factor converts moles of HCl to grams of HCl, once again using molar mass as found from the Periodic Table.

To find the mass of CO_2 produced, set up a similar dimensional analysis equation. First convert the mass of $CaCO_3$ to moles. This time, however, use the mole ratio for CO_2 and $CaCO_3$ for the second factor. Finally, use the molar mass of CO_2 to convert moles of CO_2 to grams.

$$\frac{2.59 \text{ g } CaCO_3}{1} \times \underbrace{\frac{1 \text{ mol } CaCO_3}{100.1 \text{ g } CaCO_3}}_{\uparrow \text{Factor\#1}} \times \underbrace{\frac{1 \text{ mol } CO_2}{1 \text{ mol } CaCO_3}}_{\uparrow \text{Factor\#2}} \times \underbrace{\frac{44.0 \text{ g } CO_2}{1 \text{ mol } CO_2}}_{\uparrow \text{Factor\#3}} = 1.14 \text{ g } CO_2$$

Stoichiometry: Volume-Volume Problems

In the lab, gases are often used and their volumes are easily measured. Recall that the molar volume of any gas at STP is 22.4 L. So the volume of a gas at standard temperature and pressure can be converted to moles of that gas via dimensional analysis. Combining those conversions with the mole ratios found in balanced equations, for reactions involving gases, allows for **volume-volume** stoichiometry problems to be calculated. In volume-volume problems, you are given the volume of one gas at STP and asked to determine the volume(s) of other gases involved in the reaction.

EXAMPLE

Consider the following balanced chemical reaction equation:

$$N_2 \text{ (g)} + 3H_2 \text{ (g)} \rightarrow 2NH_3 \text{ (g)}$$

To produce 0.400 L of NH_3 at STP, what volumes of nitrogen and hydrogen, also at STP, would be required?

Use dimensional analysis:

$$\frac{0.400 \text{ L } NH_3}{1} \times \underbrace{\frac{1 \text{ mol } NH_3}{22.4 \text{ L } NH_3}}_{\uparrow \text{Factor \#1}} \times \underbrace{\frac{1 \text{ mol } N_2}{2 \text{ mol } NH_3}}_{\uparrow \text{Factor \#2}} \times \underbrace{\frac{22.4 \text{ L } N_2}{1 \text{ mol } N_2}}_{\uparrow \text{Factor \#3}} = 0.200 \text{ L } N_2$$

The first factor in the dimensional analysis equation converts the volume of NH_3 to moles. The second factor uses the mole ratio from the balanced equation to convert moles of NH_3 to moles of N_2. Finally, the third factor converts moles of N_2 to the volume of N_2. Mathematically, factors #1 and #3 simply undo each other. They display a relationship of the Ideal Gas Law as well as of Avogadro's Hypothesis. Namely, there is a direct relationship between the volumes of gases, at the same temperature and pressure, and their numbers of particles (measured in moles). In other words, mole ratios can be construed as volume ratios between gases existing at the same temperature and pressure. The only factor needed to solve the previous problem mathematically was factor #2. To express the dimensional analysis equation properly, though, requires using the mole ratios as volume ratios as is done here:

$$\frac{0.400 \text{ L NH}_3}{1} \times \frac{1 \text{ L N}_2}{2 \text{ L NH}_3} = 0.200 \text{ L N}_2$$

To find the amount of H_2 required to produce the 0.400 L of NH_3 requires a similar equation but with a different ratio between the gases:

$$\frac{0.400 \text{ L NH}_3}{1} \times \frac{3 \text{ L H}_2}{2 \text{ L NH}_3} = 0.600 \text{ L H}_2$$

Using mole ratios, i.e., the coefficients from the balanced equation, as volume ratios saves time. It also makes volume-volume problems less cumbersome to solve. The relationship between the volumes of reacting gases was first noted by the French scientist Joseph Louis Gay-Lussac and is sometimes called **Gay-Lussac's Law of Combining Gases**. This law states that when only gases are involved in a chemical reaction, the volumes of the reacting gases and the volumes of the gaseous products are in small whole-number ratios with each other. Those small whole numbers are the coefficients in the balanced reaction equation.

Stoichiometry: Mass-Volume or Volume-Mass Problems

Reactions often involve gases and other phases of matter. In those reactions, it is common to know the mass of one substance involved in the chemical process and the need to determine the volume of a different substance, such as a gas. Likewise, it is not unusual to know the volume of a gas taking part in a reaction and the need to determine the mass of another substance, often a solid or liquid. Even if the reaction is taking place at STP, Gay-Lussac's Law cannot be taken advantage of here since both of the substances are not gases and the information desired is not restricted to just volumes. In other words, Gay-Lussac's Law applies only when all the substances being considered are gases.

EXAMPLE

In the reaction below, what mass of magnesium is required to produce 0.250 L of H_2 at STP?

$$Mg(s) + 2\,HCl(aq) \rightarrow MgCl_2(aq) + H_2(g)$$

The solution to this problem uses both molar mass and molar volume. Use dimensional analysis:

$$\frac{0.250 \text{ L H}_2}{1} \times \frac{1 \text{ mol H}_2}{22.4 \text{ L H}_2} \times \frac{1 \text{ mol Mg}}{1 \text{ mol H}_2} \times \frac{24.3 \text{ g Mg}}{1 \text{ mol Mg}} = 0.271 \text{ g Mg}$$

Reactions involving gases and other phases of matter NOT at STP are very common. Obviously, Gay-Lussac's Law cannot be used to solve such problems since both of the substances in question are not gases. Additionally, the relationship 22.4 L = 1 mole of the gas cannot be used since the reaction is not taking place at STP. To solve problems such as these, the Ideal Gas Law, $PV = nRT$, must be considered. The Ideal Gas Law can be manipulated to solve for the moles of gas at any temperature and pressure as long as the volume is supplied $\left(n = \frac{PV}{RT} \right)$. Likewise, the volume of a gas can be determined if the number of moles of the gas is known along with its temperature and pressure $\left(V = \frac{nRT}{P} \right)$.

Problems with an Excess of One Reactant or a Limiting Reactant

TIP

Remember this recipe analogy.

It will not always be true that the amounts given in a particular problem are exactly in the proportion required for the reaction to use up all of the reactants. In other words, at times some of one reactant will be left over after the other has been used up. This is similar to the situation in which two eggs are required to mix with one cup of flour in a particular recipe, and you have four eggs and four cups of flour.

Since two eggs require only one cup of flour, four eggs can use only two cups of flour and two cups of flour will be left over.

A chemical equation is very much like a recipe.

EXAMPLE

Consider the following reaction:

$$CH_4\,(g) + 2O_2\,(g) \rightarrow CO_2\,(g) + 2H_2O\,(g)$$

If you are given 15.0 grams of methane (CH_4) and 15.0 grams of oxygen (O_2), how many grams of carbon dioxide gas can be produced? Which reactant will be left over? How much of this reactant will not be used?

Once again, dimensional analysis will be used to solve this problem. However, two equations will be required as it will be necessary to determine how much carbon dioxide can be produced from each reactant:

$$\frac{15.0\text{ g }CH_4}{1} \times \frac{1\text{ mol }CH_4}{16.0\text{ g }CH_4} \times \frac{1\text{ mol }CO_2}{1\text{ mol }CH_4} \times \frac{44.0\text{ g }CO_2}{1\text{ mol }CO_2} = 41.3\text{ g }CO_2$$

$$\frac{15.0\text{ g }O_2}{1} \times \frac{1\text{ mol }O_2}{32.0\text{ g }O_2} \times \frac{1\text{ mol }CO_2}{2\text{ mol }O_2} \times \frac{44.0\text{ g }CO_2}{1\text{ mol }CO_2} = 10.3\text{ g }CO_2$$

Since the oxygen can produce only 10.3 g CO_2 (the lesser of the quantities of CO_2 shown above), it is referred to as the **limiting reactant**. There simply is not enough oxygen available to make what the methane has the potential to produce. That smaller amount of CO_2 is referred to as the **theoretical yield**. A portion of the CH_4 (the **reactant in excess**) will be used, however, to produce the 10.3 g of CO_2. To determine that amount, another dimensional analysis is needed.

$$\frac{15.0\text{ g }O_2}{1} \times \frac{1\text{ mol }O_2}{32.0\text{ g }O_2} \times \frac{1\text{ mol }CH_4}{2\text{ mol }O_2} \times \frac{16.0\text{ g }CH_4}{1\text{ mol }CH_4} = 3.75\text{ g }CH_4\text{ (needed)}$$

Since only 3.75 g of CH_4 is needed, the amount of CH_4 in excess can be determined by subtraction:

$$\underset{\text{(have)}}{15.0 \text{ g } CH_4} - \underset{\text{(need)}}{3.75 \text{ g } CH_4} = \underset{\text{(left over)}}{11.25 \text{ g } CH_4} \quad \text{(or 11.3 g when considering significant figures)}$$

Percent Yield of a Product

In most stoichiometric problems, we assume that the results are exactly what we would theoretically expect. In reality, the resulting theoretical yield is rarely the actual yield. Why the actual yield of a reaction may be less than the theoretical yield occurs for many reasons. Some of the product is often lost during the purification or collection process.

Chemists are usually interested in the efficiency of a reaction. The efficiency is expressed by comparing the actual and the theoretical yields.

The **percent yield** is the ratio of the actual yield to the theoretical yield, multiplied by 100%.

$$\text{percent yield} = \frac{\text{actual yield}}{\text{theoretical yield}} \times 100$$

EXAMPLE

Aluminum is commonly produced by the smelting of aluminum oxide into aluminum metal by the reaction below:

$$2 Al_2O_3 \text{ (dissolved)} + 3C(s) \rightarrow 4 Al(\ell) + 3CO_2 \text{ (g)}$$

If 3.89 kg of aluminum oxide is smelted and the actual yield of Al is 1.95 kg, what is the percent yield associated with the process?

The *theoretical yield* of aluminum from that amount of aluminum oxide can be found using dimensional analysis:

$$\frac{3,890 \text{ g } Al_2O_3}{1} \times \frac{1 \text{ mol } Al_2O_3}{102 \text{ g } Al_2O_3} \times \frac{4 \text{ mol Al}}{2 \text{ mol } Al_2O_3} \times \frac{27.0 \text{ g Al}}{1 \text{ mol Al}} = 2,060 \text{ g Al}$$

The *percent yield* can then be found:

$$\text{percent yield} = \frac{1,950 \text{ g Al (actual yield)}}{2,060 \text{ g Al (theoretical yield)}} \times 100\% = 94.7\%$$

Liquids, Solids, and Phase Changes

Liquids and solids are considered **condensed phases of matter**. By itself, this is a significant differentiation from the phase of matter called a gas. But there are many other important characteristics that distinguish solids and liquids from gases as well as those that distinguish solids from liquids. Macroscopically (i.e., in a form visible to the naked eye), there are several ways to differentiate solids, liquids, and gases. But in order to understand the states of matter well, and in addition how they change from one to another, it is important to be able to envision these phases from a mind's-eye perspective that takes into account what they are like on an atomic/molecular level.

General Characteristics of Solids, Liquids, and Gases

Macroscopically, solids are seen as rigid materials that maintain their shape independent of the container they are in. This rigidity is what differentiates them from liquids and gases, which can both be described as **fluid** because they can flow and take on the shape of their containers. Although liquids and solids are different with regard to their visible shapes, they are similar with regard to their measurable volumes. Liquids, unlike gases, do not assume the volume of their container. Gases spread out to take up the volume of the container in which they find themselves. Solids act like liquids: their volume is not dependent upon or determined by the container they occupy. Table 7.1 below summarizes these points.

TABLE 7.1 Shape and Volume Characteristics of All Three Phases

Phase:	Solid	Liquid	Gas
Shape:	Definite	Variable	Variable
Volume:	Definite	Definite	Variable

In order to explain these observable differences and similarities between the phases of matter, an understanding of the states of matter from a particle perspective is needed. Recall that liquids and solids were described above as *condensed* phases of matter. This is so because, on a particulate level, the atoms and/or molecules in a liquid or solid are in relatively close proximity. In the gas phase, on the other hand, the particles are relatively far apart. This is true whether different states of the *same* substance at *different* temperatures or different states of *different* substances at the *same* temperature are being compared. Notice how important it is that the temperature and the substance being evaluated be taken into consideration. The temperature of the material is critical because it is a measure of the average kinetic energy possessed by the substance. Kinetic energy can be described as the energy associated with the motion of a substance's particles. The substance itself is important because the attractive forces between the pertinent particles making up the substance dictate the degree to which the particles will want to cling to each other and make the phase of the substance condensed or not.

In order to encourage a substance to behave as a solid, it makes sense that its temperature (i.e., its average kinetic energy) be relatively low. This way the motion of the particles will not inhibit the particles from interacting and thereby cause them to cling to each other. Likewise, substances are encouraged to be in the solid state when the attractive force strength between the particles is relatively high. This way the particles will be able to overcome their independent motion and be made to cling to each other once again. Similarly, gases may be said to exist when the temperature is relatively high and the attractive force strength between the particles is relatively low. In these circumstances the particles can spread out without interacting with each other. The liquid state is generally present when the temperature and attractive force strength for a particular substance are in a range between that for solids and gases. Since the liquid phase is *between* the gas and solid phases, liquids are, under certain conditions, more like solids (i.e., condensed), while under other conditions, more like gases (i.e., fluid). Tables 7.2 and 7.3 summarize general occurrences of the above points for most substances.*

TABLE 7.2 States of Matter for the *Same* Substance at *Different* Temperatures

Phase:	Solid	Liquid	Gas
Particle Proximity:	Very close	Close	Far apart
Average Kinetic Energy:	Low	Moderate	High
Attractive Force Strength:	---------------- All the same** ----------------		

*Water constitutes an exception to the rule of relative particle proximity. That is, the molecules of water are closer in the liquid state than they are in the solid state. Ice floats on liquid water because ice is less dense.
**The attractive force strength is the same for all the phases because with the substance held constant, the particle interactions are unchanging.

TABLE 7.3 States of Matter for *Different* Substances at the *Same* Temperature

Phase:	Solid	Liquid	Gas
Particle Proximity:	Very close	Close	Far apart
Average Kinetic Energy:	---------------- All the same*** ----------------		
Attractive Force Strength:	High	Moderate to high	Low

***The average kinetic energy is the same for all phases because the temperature is unchanging.

It is noteworthy that at the same temperature, different substances have the same average kinetic energy despite being in different phases. If a substance has a temperature higher than absolute zero (or 0 K), all phases contain particles that are moving. The types of motion available to the particles of matter include vibration (shaking and stretching), rotation (spinning), and translation (moving from one point to another in 3-D space). Since solids generally contain particles that are very close and held in fixed positions by forces strong enough to make these substances rigid, for them the most significant type of motion realized is vibration. On the other hand, gases, which generally contain particles that are far apart and are little attracted to each other, exhibit all three modes of motion—particularly translation. These characteristics account for their ability to expand to the shape and volume of their container. Liquid particles exhibit all three kinds of motion; though translation is limited, there is enough to allow for the mixing of liquids left alone over time.

It is also worth emphasizing that temperature is directly related *not* to the *total* kinetic energy possessed by a sample of a substance but *only* to its *average* kinetic energy. Because of the condensed nature of solids and liquids, many more of their particles than of gases exist in a given amount of space. On the average, less than 1% of the space of a gas sample is occupied by gaseous particles, whereas the space taken up by particles in liquids and solids measures in the 70% range. This explains why gases are described as compressible in contrast to solids and liquids, which are generally incompressible and also exhibit a greater total kinetic energy at a given temperature. Since solids and liquids have many more particles to which the *average* kinetic energy can be assigned, their *total* kinetic energy is consequently higher.

A final but very important aspect of the differentiation between the phases of matter is the orderliness found in the arrangement of the particles in a particular phase. As has been stated, gases are composed of particles exhibiting all types of motion (particularly translation) with considerable space between the particles. Consequently, the word used to describe gases is **random**, a word that implies that a lack of order is associated with gases. The condensed solid and liquid phases are different in that they often display order, although to different degrees. The orderliness in liquids is commonly described as *short-range*, while the order in solids is commonly described as *long-range*. Liquids can be viewed as containing **clusters** of particles in which order is found within the cluster but not from cluster to cluster. Generally, clusters contain about 100 to 1,000 particles, with the constituents of clusters in constant change. The attractive forces are strong enough to generally maintain the clustery nature of the liquid but not strong enough to keep the particles in the fixed positions found in the solid state. Although there are many examples of solid substances that do not display sufficient long-range order to maintain definite, fixed positions (and so they are referred to as being **amorphous**), solids most often display significant orderliness. Such solids are described as

TIP

An increase in temperature increases the average kinetic energy of the molecules.

99

being **crystalline**. There are times when the three-dimensional arrangement involves such a large number of particles that this orderliness can be observed macroscopically and the geometry of the arrangement can be seen with the naked eye. Most other times, the solids are composed of an aggregation of smaller crystals, with these ordered domains arranged randomly; these solids are referred to as **polycrystalline**. Table 7.4 below summarizes the points made in the last few paragraphs.

TABLE 7.4 Motion and Orderliness Characteristics of All Three Phases

Phase:	Solid	Liquid	Gas
Particle Motion:	Vibration	Vibration, rotation, translation	Vibration, rotation, translation
Particle Orderliness:	Long-range (crystalline)	Short-range (clustery)	None (random)

Important Properties of Liquids

Evidence of attractive forces between particles of matter is easily found by investigating the properties of liquids. Although each property is different, the magnitude of the measured properties discussed below is fundamentally related to the strength of the interactions between the particles and provides direct evidence of their existence.

Surface Tension

Surface tension can be defined as the resistance of a liquid to create new surfaces as the result of an imbalance of attractive forces. The particles on the surface of a liquid lack an attraction upward and as such experience a net force inward; that is, they are drawn to the particles below. In other words, particles on the surface of a liquid have a "desire" to be interior particles. Liquids, therefore, have an impetus to possess the least amount of surface possible and often "bead up." The effect essentially creates a "skin" for a liquid that can support other materials that may, for reasons of higher density, normally want to sink. This sinking would expose the interior particles of the liquid and turn them into surface particles. Resistance to this occurrence is what constitutes surface tension.

Generally, the higher the attractive force strength between the particles, the higher the surface tension. This is so because a stronger net force inward will create a greater resistance to creating new surface particles. The specific intermolecular attractions that need to be evaluated for typical liquids at common conditions include **hydrogen bonding** (generally the strongest) followed by **permanent dipole-dipole** and then **temporary dipole-dipole** interactions (generally the weakest). Recall that temporary dipole-dipole interactions are also known as **London dispersion forces**.

Viscosity

Viscosity can be defined as the resistance of a liquid to flow. The ability of a liquid to flow is again fundamentally related to the strength of the intermolecular attractions between the molecules making up the substance. The stronger the attractions between the molecules, the less likely a liquid's ability to flow and the greater its viscosity.

TIP
Liquids that are more viscous flow more slowly.

Capillary Action

Capillary action, the attraction of the surface of a liquid to the surface of a solid, is a property closely related to surface tension. A liquid will rise quite high in a very narrow tube if a strong attraction exists between the liquid molecules and the molecules that make up the surface of the tube. This attraction tends to pull the liquid molecules upward along the surface against the pull of gravity. This process continues until the weight of the liquid balances the gravitational force. Capillary action can occur between water molecules and paper fiber, causing the water molecules to rise up the paper. When a water soluble ink is placed on the paper, the ink moves up the paper and separates into its various colored components. This separation occurs because the water and the paper attract the molecules of the ink components differently. These phenomena are used in the separation process of **paper chromatography**. Capillary action is at least partly responsible for the transportation of water from the roots of a plant to its leaves. The same process is responsible for the concave liquid surface, called a **meniscus**, that forms in a test tube or graduated cylinder.

Heat of Vaporization

The **heat of vaporization**, often symbolized ΔH_{vap}, is defined as the amount of heat required to vaporize a certain amount of a liquid at its boiling point. From a particle perspective, the process involves changing the clustery nature of a liquid into the random nature of a gas. As such, attractive interactions between the particles that cause the particles to cluster must be overcome to separate them into randomness. The separation of the particles does not make the particles move any faster, and so there is no increase in the kinetic energy throughout this process; the temperature remains constant. However, as energy is added to the system to overcome forces of attraction, the energy is stored by the substance, and the potential energy of the substance rises.

Once again, the magnitude of the heat of vaporization is dependent on the strength of the attractive forces, and so a higher heat of vaporization is generally associated with liquids that have a higher degree of intermolecular bond strength.

Equilibrium Vapor Pressure

Recall that the Maxwell-Boltzmann distribution of kinetic energy associated with a liquid indicates that most of the particles in a liquid possess a kinetic energy near the average value. Some, however, have values lower than the average, and others have values higher than the average. Now envision a situation where a high-energy particle finds itself on the surface of a liquid and possesses enough energy to break away from the cluster and turn into a gas. This is precisely what happens when **evaporation** occurs: as a certain fraction of the molecules in a liquid meet this threshold value, a portion of the liquid turns into **vapor** particles. It is important to note that this can happen below the boiling point of the liquid (where that process normally occurs) but only on its surface.

Since evaporation involves the loss of the high-energy particles from the liquid, it is commonly viewed as a **cooling process**. As the high-energy particles leave, the temperature of the liquid decreases, and the average kinetic energy necessarily goes down. The validity of this claim is easily recognized: consider how humans naturally sweat to cool themselves.

If a liquid is placed in a closed container and the temperature is held constant, evaporation will occur, and vapor particles will begin to fill the space above the liquid surface. Because the container is closed as time goes by, some of the vapor particles will have an opportunity to collide with the surface of the liquid; in the process they will perhaps lose a portion of their energy and be "captured" by the clusters of the liquid. This is precisely what happens during the process of **condensation** (*when a gas turns into a liquid*). Initially, the rate of evaporation is higher than that of condensation because of the fact that not many vapor particles are around. But as the number of vapor particles increases, so does the rate of condensation. It should now be easy to envision a point in time where the rates of evaporation and condensation are equal to each other and the number of vapor particles above the liquid becomes constant (see Figure 7.1). Situations like this, where two opposing processes are equal to each other in rate, are common in nature and are referred to as a **dynamic equilibrium**. Translating gaseous particles create pressure as the particles collide with their container walls. Consequently, the phenomenon described above is the reason why an **equilibrium vapor pressure** is associated with a given liquid.

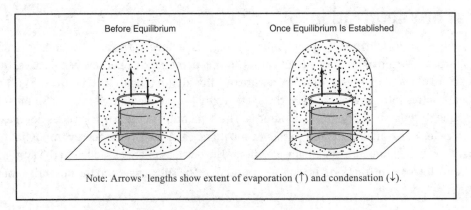

Before Equilibrium   Once Equilibrium Is Established

Note: Arrows' lengths show extent of evaporation (↑) and condensation (↓).

FIGURE 7.1 A Closed System in Dynamic Equilibrium

As we have seen with regard to the other properties of liquids, the equilibrium vapor pressure of a liquid is also fundamentally related to the attractive force strength between the particles in the liquid—this time, however, in the opposite way. Higher attractive force strengths produce lower vapor pressures. This is true because a smaller fraction of high-energy particles in a given sample will possess the threshold energy needed to break away and turn into a gas on the surface of the liquid if the attractive force holding them to the cluster is stronger. The consequence is that equilibrium will be established with a smaller number of vapor particles above the liquid and hence a lower vapor pressure.

Boiling Point

TIP

Boiling point is defined as the temperature at which the liquid's vapor pressure equals the atmospheric pressure.

Although the phase change associated with boiling (a liquid turning into a gas) is the same as seen with evaporation, boiling has several significant differences. As your own experience is likely to have verified, boiling occurs throughout the body of a liquid, not only on the surface, as happens with evaporation. This is shown by the presence of bubbles in the liquid during boiling. The requirement of bubble production creates another difference between boiling and evaporation. The creation of a bubble mandates that the vapor pressure of the gas making the bubble be high enough to withstand the pressure of the atmosphere above the liquid pushing down on the bubble to collapse it or inhibit its production altogether. Consequently, the temperature must be high enough to produce a vapor pressure for the liquid that is at least equal to the atmospheric pressure. This means that boiling occurs at a particular temperature, called the **boiling point**, where the vapor pressure of the liquid and the atmospheric pressure are the same. The ramification of that concept is that the boiling point of a liquid is variable: it is dependent on the atmospheric pressure under which the liquid exists. That is, if the atmospheric pressure on a liquid is lower (as it is above sea level, as may be seen in the mountains), the boiling point of the liquid is lower. This is so because the temperature to which the liquid must be raised to cause the vapor pressure of the liquid to be equal to the atmosphere will not be as high. The opposite is true for a liquid below sea level. The **normal boiling point** is the temperature at which the vapor pressure of a liquid is equal to 1.0 atmosphere (760 mm of Hg), the atmospheric pressure at sea level. Table 7.5 below summarizes the differences and similarities between boiling and evaporation.

TABLE 7.5 The Similarities and Differences Between Evaporation and Boiling

Process:	Evaporation	Boiling
Similarities:	Particles break away from liquid clusters to form a gas. The process is endothermic (requires energy).	
Differences:	Below the boiling point	At the boiling point
	Occurs on the surface of the liquid	Occurs throughout the body of the liquid (bubbles)

Since vapor pressure and boiling point are related to each other and vapor pressure is related to attractive force strength, boiling point must also be related to attractive force strength. Generally, higher boiling points indicate higher attractive force strength between the particles of a liquid. This makes sense: stronger bonds will require higher temperatures to make the vapor pressure high enough to equal that of the atmosphere.

Critical Temperature

There is a limit to the attractive force that binds particles together in a liquid. It is only so strong. As a result, one can envision a temperature above which the kinetic energy of the particles will always be high enough to inhibit the clustering of gas particles into a liquid. This temperature, referred to as the **critical temperature**, represents a value above which the substance cannot be condensed. Since the attractive force limit dictates the temperature at which this occurs for various substances, one sees again that the magnitude of the attractive forces must dictate the magnitude of the critical temperature. Higher intermolecular bond strengths are associated with higher critical temperatures.

Important Properties of Solids

Just as the attractive forces between the particles in liquids dictate the magnitude of the properties of liquids, the attractive forces between the particles in a solid do the same.

Heat of Fusion

The **heat of fusion**, often symbolized ΔH_{fus}, is defined as the amount of heat *required* to liquefy a certain amount of a solid at its melting point. As with heat of vaporization, the energy is used to overcome the attractive forces between the particles, not to make them move more. Consequently, the crystalline nature of a solid is transformed into the clustery nature of a liquid, and the temperature of the material does not rise. As seen previously, since the temperature is not on the rise, neither is the kinetic energy. What does increase is the potential energy of the material as the energy provided to the system is stored in the liquid phase. It is important to recognize that as a liquid is frozen (i.e., the opposite process of melting), this same quantity of energy is *released*. As is seen with heat of vaporization, here, too, higher attractive force strengths result in a higher heat of fusion for a given substance.

Melting (and Freezing) Point

Earlier, the boiling point of a liquid was shown to depend upon the vapor pressure of the liquid. The melting and freezing points of a substance show a similar dependency although for a different reason. Just like liquids, solids exhibit an equilibrium vapor pressure if placed in a closed container. High-energy particles on the surface of the solid can break away from the crystal and enter the gaseous state. The process of a solid turning into a gas is called **sublimation**. When the rate of sublimation equals the rate of **vapor deposition** (the opposite of sublimation), equilibrium is established and the number of gaseous particles above the solid is constant, creating a certain pressure. Also, as with the liquid, the vapor pressure of the solid is dependent on the temperature of the material. Consequently, there is a temperature at which the vapor pressure of the solid is equal to the vapor pressure of the liquid. This temperature at which melting (or freezing) takes place is called the **melting** (or **freezing**) **point**, depending on whether energy is being added to or taken away from the system. The temperature at which this condition is met under 1 atmosphere of pressure is called the *normal* melting (or freezing) point. Since the vapor pressure of a substance is contingent upon the strength of the attractive forces between the particles, and the melting/freezing point is dictated by the liquid's and solid's vapor pressure, then the melting/freezing point depends on the strength of the attractive forces. Higher attractive force strength is indicative of a higher melting/freezing point.

Electrical Conductivity

In order for a substance to be electrically conductive, it must contain charged particles that can move translationally. For most types of solids, this is not practically possible because the particles, even if they are charged, are held in fixed positions. The only exception is for metallic solids in which the core of the metal atoms is held reasonably firm by the *metallic bond* but the valence electrons of those atoms are not. Recall how the metallic bond can be described as a *mobile* "sea of electrons" surrounding the remaining positive cores of the atoms.

Solubility

The solubility of solids varies with the type of bonding between the particles making up the solid and the kind of substance in which the solid will be dissolved. Although a more in-depth discussion on the solution process will be seen later in this chapter, a good rule of thumb concerning the degree to which a solute (what is being dissolved) will dissolve in a solvent (what is being dissolved into) is the phrase "*like dissolves like.*" Keep in mind that this is a *general rule* and that, for good reason, many exceptions exist. The "like dissolves like" rule refers to the nature of the materials involved with reference to their distribution of charged parts. For example, solids that have permanent charges (i.e., ionic or polar molecular substances) tend to dissolve in similar substances (i.e., polar molecular liquids). Likewise, nonpolar molecular solids tend to dissolve in nonpolar molecular liquids.

Types of Solids

As seen in the many examples describing the properties of solids, the particles that make up a solid and the types of bonds with which they are held together are critical pieces of information needed in predicting the relative behavior of a material. Accordingly, solids can be categorized based on those characteristics. Table 7.6 below summarizes the different types of solids and the properties generally associated with them.

TABLE 7.6 Types of Solids

Type of Solid	Examples	Pertinent Particles	Type of Bonds	General Properties
Macromolecular (Also known as a network solid)	C (as in diamond or graphite), SiC, SiO_2	Nonmetal atoms	Covalent	High heat of fusion and melting point; low electrical conductivity and solubility in water
Ionic	NaCl, $MgSO_4$, $K_2Cr_2O_7$	Metal and non-metal ions	Ionic	High heat of fusion, melting point, and solubility in water*; low electrical conductivity
Metallic	Cu, Mg, Hg	Metal atoms	Metallic	Moderate to high heat of fusion and melting point; high electrical conductivity; low solubility in most solvents
Molecular	SO_2, NH_3, CH_4	Molecules	Intermolecular	Low heat of fusion, melting point, and electrical conductivity; variable solubility in water

*There are many ionic substances insoluble in water.

There is a special type of solid that contains ions *and* molecules combined in a single substance; it is called a **hydrate**. Hydrates can be viewed as containing fully charged ions and polar water molecules surrounded by and attracted to each other in a crystal structure that appears dry despite the presence of the water. A hydrate's formula displays the formula for the regular ionic substance followed by the number of water molecules associated with it per formula unit. Examples include $CuSO_4 \cdot 5H_2O$ and $BaCl_2 \cdot 2H_2O$ (the • in the formula is read as "with"). Often, the heating of hydrated crystals releases their water of hydration and alters the crystal structure of the remaining "anhydrous" ionic material; this in turn causes a color change in the solid substance. Conversely, other ionic solids are so attracted to water molecules that they readily *become* hydrates by absorbing moisture from the air; these are referred to as **hygroscopic** materials.

Phase Diagrams

The previously described properties of solids, liquids, and gases, as well as the processes relating to their change from one to another, all point toward the importance of temperature and pressure in determining the stable state of matter for a given substance. A **phase diagram** provides a graphical way to summarize the conditions of those parameters that dictate the phase the substance will primarily find itself in once equilibrium is established. Figure 7.2 is an example of a phase diagram of the substance water. It is important to note that many phase diagrams (like the one shown in Figure 7.2) are not shown to scale but are meant to convey important information about a substance. Analysis of Figure 7.2 shows that line *BD* is essentially the vapor pressure curve for water's liquid phase. Notice that when the pressure on a sample of water is 760 mm of Hg, the vapor pressure of the water matches that at 100°C, and the water will boil. However, if the pressure is raised, the boiling point temperature increases, and if the pressure is less than 760 mm of Hg, the boiling point decreases along the *BD* curve down to point *B*.

> **TIP**
>
> The *triple point* is the only temperature and pressure at which all three phases of a substance can exist.

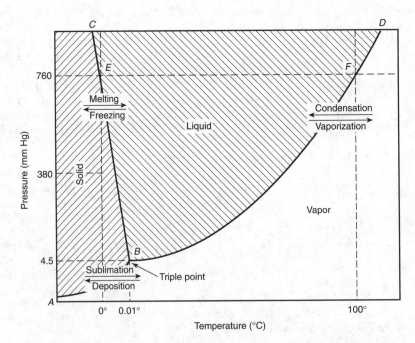

FIGURE 7.2 Partial Phase Diagram for Water (distorted somewhat to distinguish the triple point from the freezing point)

As line *BD* is essentially the vapor pressure curve for liquid water, line *AB* is the vapor pressure curve for solid water. Point *B* is one position where the vapor pressure (or VP) of the solid is equal to the vapor pressure of the liquid. As the pressure is increased, the temperature at which this remains true goes down, as seen by the negative slope of the line emanating upward from point *B*. This differs from most other substances, which display a positive slope in this condition, and has to do with the density of solid water (ice) being less than the density of liquid water (discussed earlier in this chapter). Recall too that when the VP of a solid is equal to the VP of a liquid, melting and freezing take place in equilibrium with each other. Consequently, the normal melting/freezing point for water (i.e., at 760 mm of Hg) is shown on the diagram to be 0.0°C. Again, this point is affected by pressure along line *BC* so that if the pressure is decreased, the melting/freezing point is slightly higher up to point *B*, or 0.01°C.

Point *B* on the diagram is unique in that not only are melting and freezing in equilibrium with each other but so are *all* other phase changes. Point *B*, commonly referred to as the **triple point**, describes the only combination of temperature and pressure where all three phases are simultaneously stable. Combinations of temperature and pressure that fall on the lines in these diagrams describe stable two-phase regions. All other combinations represent stable single-phase regions for the substance.

Another noteworthy point on a phase diagram is the end of the vapor pressure curve for the liquid. The phase diagram for water in Figure 7.2 cannot show this because of the figure's scale, but it is shown on many other diagrams. As earlier discussed in reference to the properties of liquids, this point, called the critical temperature, is the temperature above which a substance cannot exist as a liquid. That explains why the vapor pressure curve comes to an end. If the liquid cannot exist, there can be no vapor pressure measured for it. Note that if a substance is at its critical temperature, it can exist as a liquid, but in order to be so, it must also be at a high enough pressure. The pressure needed to liquefy a gas at its critical temperature is referred to as its **critical pressure**. The combination of critical temperature and critical pressure, often seen in phase diagrams, is called its **critical point**.

Solutions

A solution is generally considered a homogeneous physical mixture of substances, a solute and a solvent. Understanding that combination from the perspective of phase and phase change provides a deep understanding of the process of solution making and the properties of a solution once made.

Since there are three phases possible for both the solute and the solvent, nine possible kinds of solutions exist from a phase perspective. All may be discussed in beginning chemistry courses, but the two most common are solids dissolved in liquids and gases dissolved in liquids. Since solutions must be homogeneous and consequently only one phase, the material being dissolved (in the two examples given) must change phase as the solution is formed; this material is the **solute**. The substance not changing phase is the **solvent**. In other examples, if there is no phase change (as when a liquid is dissolved in a liquid), the substance in the smaller amount is generally considered the solute.

Water (a Common Solvent)

Water is so often involved in chemistry that it is important to have a rather complete understanding of this compound and its properties. Pure water has become a matter of national concern. Although commercial methods of purification will not be discussed here, the usual laboratory method of obtaining pure water, distillation, will be covered.

Purification of Water

The process of distillation involves the evaporation and condensation of the water molecules. The usual apparatus for the distillation of any liquid is shown in Figure 7.3.

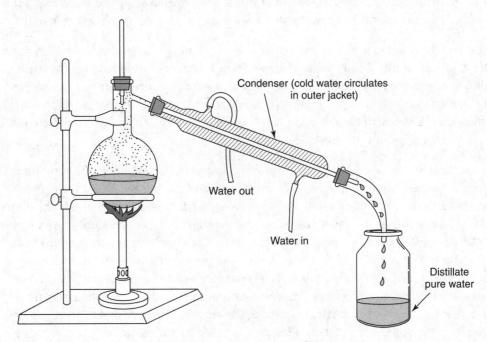

Condenser (cold water circulates in outer jacket)

Water out

Water in

Distillate pure water

FIGURE 7.3 Distillation of Water

TIP

In distillation, first boil and then condense.

This method of purification will remove any substance that has a boiling point higher than that of water. It cannot remove dissolved gases or liquids that boil off before water. These substances will be carried over into the condenser and subsequently into the distillate.

Polarity and Hydrogen Bonding in Water

Water is different from most liquids in that it reaches its greatest density at 4°C and then its volume begins to expand. By the time water freezes at 0°C, its volume has expanded by about 9%. Most other liquids contract as they cool and change state to a solid because their molecules have less energy, move more slowly, and are closer together. This abnormal behavior of water can be explained as follows. X-ray studies of ice crystals show that H_2O molecules are bound into large molecules in which each oxygen atom is connected through **hydrogen bonds** to four other oxygen atoms as shown in Figure 7.4.

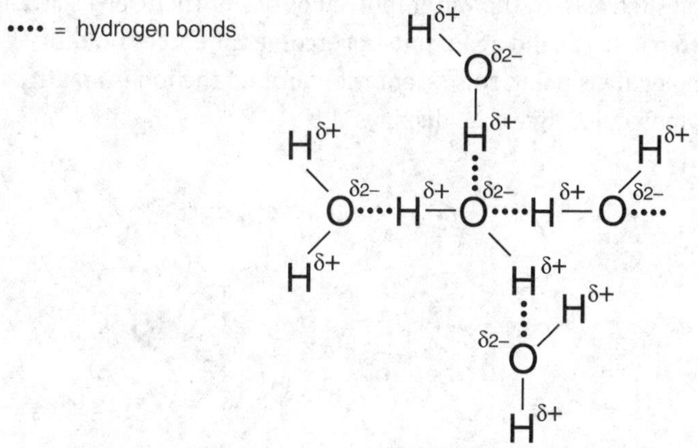

•••• = hydrogen bonds

FIGURE 7.4 Study of Ice Crystal

This rather wide open structure accounts for the low density of ice. As heat is applied and melting begins, this structure begins to collapse, but not all the hydrogen bonds are broken. The collapsing increases the density of the water, but the remaining bonds keep the structure from completely collapsing. As heat is absorbed, the kinetic energy of the molecules breaks more of these bonds as the temperature rises from 0° to 4°C. At the same time, this added kinetic energy tends to distribute the molecules farther apart. At 4°C, these opposing forces are in balance—thus resulting in the greatest density. Above 4°C, the increasing molecular motion again causes a decrease in density since it is the dominant force and offsets the breaking of any more hydrogen bonds.

This behavior of water can be explained by studying the water molecule itself. The water molecule is composed of two hydrogen atoms bonded by a polar covalent bond to one oxygen atom as seen in Figure 7.5.

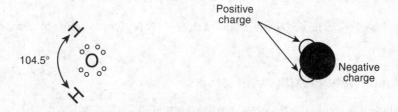

FIGURE 7.5 Representations of a Water Molecule

Because of the polar nature of the bond, the molecule exhibits the charges shown in the above drawing. It is this polar charge that causes polar bonding as the hydrogen bonds. This bonding is stronger than the usual molecular attraction called van der Waals forces or dipole-dipole attractions.

Water Solutions and Mixtures

To make molecules or ions of another substance go into solution, water molecules must overcome the forces that hold these molecules or ions together. The mechanism of the actual process is complex. To make sugar molecules go into solution, the water molecules cluster around the sugar molecules, pull them off, and disperse, forming the solution.

For an ionic crystal such as salt, the water molecules orient themselves around the ions (which are charged particles) and again must overcome the forces holding the ions together. Since the water molecule is polar, this orientation around the ion is an attraction of the polar ends of the water molecule. Figure 7.6 displays this:

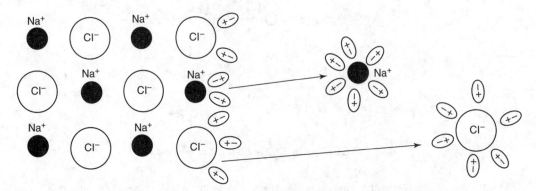

FIGURE 7.6 Water Dissolving Sodium Chloride

Once surrounded, the ion is insulated to an extent from other ions in solution because of the dipole property of water. The water molecules that surround the ion differ in number for various ions, and the whole group is called a **hydrated ion**.

In general, as stated in the preceding section, polar substances and ions dissolve in polar solvents and nonpolar substances such as fats dissolve in nonpolar solvents such as gasoline. The process of going into solution is **exothermic** if energy is released in the process, and **endothermic** if energy from the water is used up to a greater extent than energy is released in freeing the particle.

When two liquids are mixed and they dissolve in each other, they are said to be completely **miscible**. If they separate and do not mix, they are said to be **immiscible**.

Two molten metals may be mixed and allowed to cool. This gives a "solid solution" called an **alloy**.

Continuum of Water Mixtures

Figure 7.7 shows the general sizes of the particles found in a water mixture.

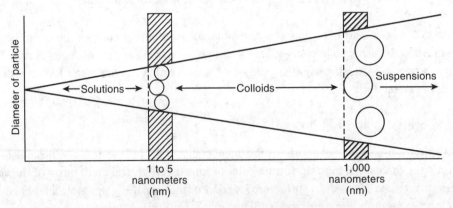

FIGURE 7.7 Size of Particles in Water Mixture

The basic difference between a colloid and a suspension is the diameter of the particles dispersed. All the boundaries marked in Figure 7.7 indicate only the general ranges in which the distinctions between solutions, colloids, and suspensions are usually made.

The characteristics of water mixtures are given in Table 7.7:

TABLE 7.7 The Characteristics of Water Mixtures

Solutions	Colloids	Suspensions
. . . . 1 nm . . . . .1,000 nm		
Clear; may have color Particles do not settle.		Cloudy; opaque color Settle on standing
Particles pass through ordinary filter paper.		Do not pass through ordinary filter paper
Particles pass through membranes.	Do not pass through semipermeable membranes such as animal bladders, cellophane, and parchment, which have very small pores*	
Particles are not visible.	Visible in ultramicroscope	Visible with microscope or naked eye
	Show Brownian movement	No Brownian movement

Separation of a solution from a colloidal dispersion through a semipermeable membrane is called dialysis.

When a bright light is directed at right angles to the stage of an ultramicroscope, the individual reflections of colloidal particles can be observed to be following a random zigzag path. This is explained as follows: The molecules in the dispersing medium are in motion and continuously bumping into the colloidal particles, causing them to change direction in a random fashion. This motion is called **Brownian movement** after the Scottish botanist Robert Brown, who first observed it.

Solubility

The degree to which a solute can dissolve in a solvent, called **solubility**, varies with the solute/solvent combination, as seen previously (recall "like dissolves like"). Additionally, temperature influences the solubility of a solute in a solvent. This should make sense, as the production of a solution often involves the solute changing phase and phase change is influenced by temperature, as previously noted. Figure 7.8 displays the solubility of several ionic solids in liquid water as a function of temperature.

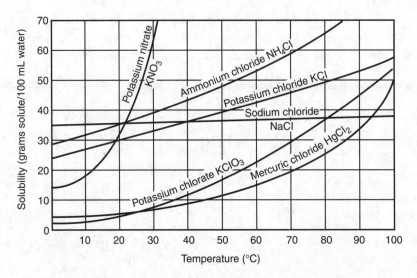

FIGURE 7.8 Solubility Curves of Some Common Salts

Although it is most common for the solubility of solids dissolved in liquids to increase with temperature, the solubility of gases varies oppositely. Gases dissolve to a greater degree in a liquid as the temperature decreases. Recognition that a major step in the process of dissolving a gas solute in a liquid solvent requires the gas to condense (remember that solutes are the substances that change phase) suggests that lowering the temperature would aid in this process. In the same light, raising the pressure of a gaseous solute above a liquid solvent would also increase the solubility of the gas in the liquid as condensation is encouraged when the pressure of the gas is increased. Since solids and liquids are already condensed phases of matter, changes in pressure have little impact on the solubility of a solid in a liquid. Table 7.8 below summarizes these solubility trends.

TABLE 7.8 Solubility Trends

Phases Involved:	Temperature Effect	Pressure Effect
Solid dissolved in a liquid	Solubility generally increases with temperature.	No significant influence on solubility.
Gas dissolved in a liquid	Solubility generally decreases with temperature.	Solubility generally increases as the pressure of the gas increases.

Factors That Affect Rate of Solution Making (How Fast They Go into Solution)

The following procedures increase the rate of solution making.

Pulverizing   increases surface exposed to solvent.

Stirring   brings more solvent that is unsaturated into contact with solute.

Heating   increases molecular action and gives rise to mixing by convection currents. (This heating affects the solubility as well as the rate of solution making.)

Semi-Quantitative Expressions of Concentration

The solubility of a substance displayed by a solubility curve describes the maximum amount of solute that can be dissolved in a given amount of solvent at a particular temperature. Since solutions are fundamentally physical mixtures, there is no set ratio concerning the amount of solute and solvent in a solution. The solubility of a substance is recognition, however, that (in most cases) there are limits. When less than the amount of solute that can be dissolved in the solvent is dissolved, the solution is said to be **unsaturated**. Similarly, when the maximum amount of solute that can be dissolved in the solvent is dissolved, the solution is called **saturated**. Interestingly, there are, for certain solute/solvent combinations, situations where more than the typical amount of solute that can be dissolved in the solvent is dissolved; such a solution is described as **supersaturated**. With solids dissolved in liquids, this situation can occur in a quasi-stable manner when solvents are combined with more solute than can normally be dissolved and heated to get the undissolved solute to go into solution. Upon cooling, the "extra" solute can remain in solution, although somewhat tenuously.

Independent of the solubility of solute and solvent, the term **dilute** expresses that a small amount of solute is dissolved in the solvent for a given solution. When a large amount of solute is dissolved, the solution is described as **concentrated**. The exact amount of solute dissolved in each case is not specified, but chemists can easily apply appropriate terms to solutions as they gain experience in solution making and usage. It is common for inexperienced chemists to confuse dilute with unsaturated as well as concentrated with saturated, but they are not the same. A solution can be both saturated and dilute if the solubility of the solute in the solvent is low. If the solubility is low, large amounts of solute cannot be dissolved, and the solution is necessarily dilute even though the maximum amount of dissolvable solute has been reached. Conversely, solutes with high solubility in a solvent may not reach the maximum amount of dissolvable solute but still contain large amounts of solute and hence be described as both unsaturated and concentrated.

The term **soluble** is used to describe a solute when a reasonable amount of solute can dissolve in a solvent. **Insoluble** means this is not the case. Unlike the terms "dilute" and "concentrated," "soluble" and "insoluble" have a benchmark that differentiates them. Chemists generally recognize a solute as soluble if 0.10 moles of the solute can be dissolved in the amount of solvent needed to make 1.0 liter of solution.

Rules for Solubility (Concerning Ionic Solids in Water at Room Temperature)

Compounds containing:

- nitrate, acetate, or chlorate are generally **soluble**.
- sodium, potassium, and ammonium are generally **soluble**.
- chlorides are generally **soluble**, with noteworthy exceptions being those also containing silver, mercury (I), and lead (II).
- sulfates are generally **soluble**, with noteworthy exceptions being those also containing lead (II), barium, strontium, and calcium.
- carbonate, phosphate, silicate, and sulfide are generally **insoluble**, with noteworthy exceptions being those also containing sodium, potassium, and ammonium.
- hydroxides are generally **insoluble**, with noteworthy exceptions being those also containing sodium, potassium, ammonium, calcium, barium, and strontium.

> **TIP**
> You should be familiar with these general rules of the solubility of solids.

Quantitative Expressions of Concentration

The semi-quantitative terms referring to the amounts of solution parts present in a solution lack the precise information chemists often need about them. There are ways, however, to describe particularly the amounts of those solution parts useful in common chemical calculations.

Percent by mass considers the masses of the solution parts. With reference to the solute, it is defined as:

$$\frac{\text{Grams of solute}}{\text{Grams of solution}} \times 100\% = \text{percent by mass (solute)}$$

EXAMPLE

At 25°C, a saturated solution of potassium nitrate contains 45 grams of KNO_3 dissolved in 100.0 grams of water. What is the percent by mass (with regards to KNO_3) in this solution?

To solve this problem one needs to recognize that KNO_3 is the solute and has a mass of 45 grams and the mass of the solution (both solute and solvent) is 145 grams.

$$\frac{45 \text{ g}}{145 \text{ g}} \times 100\% = 31\%$$

Molarity (abbreviated M) considers the moles of solute dissolved and the volume of the overall solution; it is defined as:

$$\frac{\text{Moles of solute}}{\text{Liter of solution}} = \text{Molarity}$$

EXAMPLE

What mass of KNO_3 would be needed to produce 250 mL of solution with a concentration of 0.500 M, and how would this solution be produced in the laboratory?

This is a common calculation chemists do when solving problems associated with solutions. A rearrangement of the molarity equation above yields:

$$\text{Molarity} \times \text{Volume (in liters)} = \text{Moles of solute}$$

Once moles of solute are determined, the amount must be converted into grams via dimensional analysis.

$$0.500 \text{ M} \times 0.250 \text{ L} = 0.125 \text{ moles of solute}$$

$$\frac{0.125 \text{ moles}}{1} \times \frac{101 \text{ grams}}{1 \text{ mole}} = 12.6 \text{ grams}$$

To produce this solution, 12.6 grams of KNO_3 would be placed in a **volumetric flask** and enough water added to bring the solution to a volume of 250 mL. A volumetric flask is a special device commonly used in the laboratory to make solutions. Precisely etched in the neck of the flask is *only one* graduation, which defines the volume of the solution in the container.

Dilution

Concentrated solutions are often bought or made as stock solutions in chemical laboratories for commonly dissolved substances. Solutions that are less concentrated (or more dilute) than those are also often needed, and so problems associated with dilution are often encountered. Solving a dilution problem, once again, takes advantage of the molarity formula as rearranged above, where molarity times volume equals the number of moles of solute in the solution. To solve a dilution problem, one simply needs to keep in mind that the moles of solute received from the more concentrated solution stay the same in the diluted solution; they simply exist in a greater volume. From this it should be easy to rationalize the dilution equation:

$$M_c V_c = M_d V_d$$

where M_c and V_c represent the molarity and volume in the concentrated case and M_d and V_d represent the molarity and volume in the diluted case. In either case, the number of moles remains the same (i.e., is equal) because all that is involved in dilution is taking a certain volume of the concentrated solution and adding some water to it to make it less concentrated. In other words, if the molarity goes down, the volume must go up to keep the number of moles the same. The equation is particularly useful because typically the molarity and volume for the desired, less concentrated solution is known, as is the molarity of the more concentrated solution. Therefore, the volume of original solution to be diluted is easily found and placed in a volumetric flask to which water is added to complete the dilution.

EXAMPLE

A solution of hydrochloric acid (HCl dissolved in H_2O) is commonly purchased in large quantities at a concentration where it is saturated, 12.1 M. Solutions of less than 12.1 M are typically used, however. Describe how to make 0.500 L of solution of 3.00 M HCl from the 12.1 M stock solution.

Values for three of the four variables found in the dilution equation are provided in the problem:

$$M_c = 12.1 \text{ M} \qquad V_c = ? \qquad M_d = 3.00 \text{ M} \qquad V_d = 0.500 \text{ L}$$

Plugging these values into the equation gives:

$$12.1 \text{ M} \times V_c = 3.00 \text{ M} \times 0.500 \text{ L}$$

Solving for the volume of the original (more concentrated) solution needed gives:

$$V_c = 0.124 \text{ L}$$

Therefore, in order to produce the diluted solution, 0.124 L (i.e., 124 mL) of the concentrated HCl solution should be transferred via a pipette to a 0.500 L (i.e., 500 mL) volumetric flask and enough water added to bring the solution up to the line on the flask.

Colligative Properties of Solutions

Properties of solutions that show predictable variation from those of the solvent itself and that are dependent primarily on the number of solute particles present in the solution (not particularly on the identity of those particles) are called **colligative properties**.

Vapor pressure is one of those properties. Whenever a solute is dissolved in a solvent, new interactions between the particles present ensue. Generally, in the case of a solid or liquid solute dissolved in a liquid solvent, the interactions are such that the solvent is inhibited from evaporating to the same degree it had in its pure state, and so the vapor pressure of the solution is lower than that of the pure solvent. This phenomenon, called **vapor pressure lowering**, is displayed graphically below as a portion of a phase diagram that has been modified to show the vapor pressure versus temperature for *both* pure water and a solution in which water is the solvent. Notice that the vapor pressure increases as the temperature does for both solution and pure solvent, but at any given temperature the vapor pressure of the solution is lower than that of the pure water. If you recall that boiling occurs when the vapor pressure of a liquid is equal to the atmospheric pressure, it becomes evident that the boiling point of a solution will be different from the boiling point of the pure liquid solvent because the vapor pressure has been lowered. This is also shown graphically below in Figure 7.9. The higher temperature required to make the vapor pressure of the solution equal to the atmospheric pressure (normally 760 mm of Hg) causes the boiling point to be raised. This phenomenon, called **boiling point elevation**, is commonly taken advantage of to get liquids to boil at higher temperatures. Since foods cook faster when the water they are cooked in is hotter, chefs often put salt into the water to raise its temperature.

Vapor Pressure versus Temperature for Water and a Solution

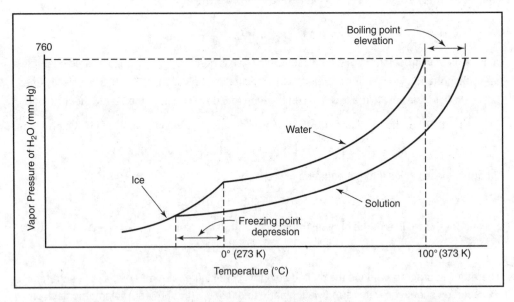

FIGURE 7.9 Vapor Pressure Curves for Water and an Aqueous Solution

In addition to elevating the boiling point of a liquid solvent, added solutes change the freezing point of the solvent as well. This is also seen graphically above. Freezing occurs when the vapor pressure of a solid is equal to the vapor pressure of the liquid with which it is associated. Since the vapor pressure of the liquid solution is lowered relative to that of the pure solvent, the liquid will not freeze at the normal freezing point because the liquid and the solid no longer share a vapor pressure value at that temperature. Instead, the temperature must be lowered for this to be true. Upon lowering the temperature, the vapor pressure lowers in both the solid and liquid solution, but the solid's vapor pressure lowers to a greater extent with each degree drop in temperature. Consequently, there will be a temperature at which the vapor pressures are again equal and a new, lower freezing point is realized. One common practice that takes advantage of this phenomenon, called **freezing point depression**, is the wintertime use of salt to melt ice on streets and roads.

As previously mentioned, the extent to which solutes lower the vapor pressure, elevate the boiling point, and depress the freezing point of solvents is dependent simply on the number of particles dissolved in the solvent, not on the particular identity of the solute particles found in the solution. But the nature of the solute does make a difference in terms of the relative degree to which the effect is noticed. This is due to the fact that some solutes dissolve and remain principally intact as the particle they were before the dissolution process. Others dissociate upon dissolution and produce a larger number of particles having a greater impact on the colligative properties.

Most molecular solutes that dissolve in a solvent will not dissociate into multiple particles, and so the molecules remain whole and homogeneously mixed with the solvent particles upon dissolving. In the case of water as a solvent, most polar molecules tend to dissolve well but don't break apart to an appreciable degree. This means that the vapor pressure of the solution will be less modified from the pure water compared to when an equal number of formula units of an ionic substance might dissolve in the same amount of water. This is so because the formula units of ionic substances break apart into the individual ions that make up the formula unit when the ionic substance is dissolved.

EXAMPLE

Solutions made from 0.50 moles of both table sugar (sucrose) and table salt (sodium chloride) dissolved in 1.0 kg of water will exhibit higher boiling points than the water itself. Will both solutions have the same boiling point, and if not, which of the solutions will exhibit a higher boiling point?

Despite the fact that an equal number of moles of each solid are being dissolved in the same mass of water, the boiling points of these solutions will not be the same. While sucrose $(C_{12}H_{22}O_{11})$ is a molecular substance that dissolves in water because it is a polar molecule, it does not dissociate into ions. Therefore, there will be 0.50 moles of sucrose molecules dissolved in the 1.0 kg of water once the solution is made. Sodium chloride (NaCl) is an ionic substance that not only dissolves in water but, as a part of its solution process, dissociates into two ions, Na^+ and Cl^-. Therefore, there will be 1.0 mole of ions (0.50×2) present in the 1.0 kg of water once the solution is made. Since the salt solution has a larger number of particles dissolved in the same mass of water, the salt solution will have a higher boiling point. Colligative properties are independent of the identity of the dissolved particles and are dependent only on their total number.

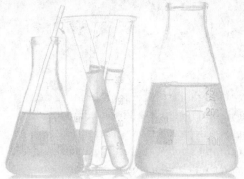

CHAPTER 8

Chemical Reactions and Thermochemistry

Learning Objectives

In this chapter, you will learn how to:

- Identify the driving force for these four major types of chemical reactions, and write balanced equations for each: combination (or synthesis), decomposition, single replacement, and double replacement.
- Identify and explain graphically enthalpy changes in exothermic and endothermic reactions.
- Solve calorimetry/heating curve problems.
- Use Hess's Law to show the additivity of heats of reactions.

Many of the kinds of reactions you may encounter in a first-year chemistry course can be placed into four basic categories: synthesis, decomposition, single replacement, and double replacement.

The first type, **synthesis**, is also known as **direct combination**. A synthesis reaction involves the making of a more complex product from the union of two or more simpler substances. A *particular* type of synthesis is a **formation reaction**, in which one mole of a compound is made; that is, a more complex product is formed from the direct combination of its elements, the elements being simpler substances in the manner in which they standardly exist. All formation reactions are synthesis reactions, but not all synthesis reactions are formation reactions. Some examples of synthesis reactions (which also happen to be formation reactions) are:

$$Zn(s) + S(s) \rightarrow ZnS(s)$$
$$Ca(s) + Cl_2(g) \rightarrow CaCl_2(g)$$
$$C(s) + O_2(g) \rightarrow CO_2(g)$$

The second type of reaction, **decomposition**, can also be referred to as **analysis**. This means the breakdown of one reactant, often a compound, to yield two or more products. These products can be simpler compounds or individual elements. In this regard, a decomposition reaction can be viewed as the reverse of a synthesis reaction. Some examples of this type are:

$$2H_2O(\ell) \rightarrow 2H_2(g) + O_2(g) \text{ (electrolysis of water)}$$
$$C_{12}H_{22}O_{11}(s) \rightarrow 12C(s) + 11H_2O(\ell)$$
$$2HgO(s) \rightarrow 2Hg(s) + O_2(g)$$

119

The third type of reaction is called **single replacement** or **single displacement**. In this type, one reactant (an element) is exchanged with an ion in a second reactant (an ionic compound). There are two types of single replacement reactions, the type depending on whether the elemental reactant is a metal or a nonmetal. The first two examples below represent *metal* single replacement reactions, and the third represents a *nonmetal* single replacement reaction. Some examples are:

$$Fe(s) + CuSO_4(aq) \rightarrow FeSO_4(aq) + Cu(s)$$
$$Zn(s) + H_2SO_4(aq) \rightarrow ZnSO_4(aq) + H_2(g)$$
$$Cl_2(g) + 2NaBr(aq) \rightarrow 2NaCl(aq) + Br_2(\ell)$$

The last type of reaction is called **double replacement** or **double displacement** because there is an actual exchange of "partners" to form new compounds. Some examples are:

$$AgNO_3(aq) + NaCL(aq) \rightarrow AgCl(s) + NaNO_3(aq)$$
$$H_2SO_4(aq) + NaOH(aq) \rightarrow Na_2SO_4(aq) + 2H_2O(\ell) \text{ (neutralization)}$$
$$BaCl_2(aq) + Na_2SO_4(aq) \rightarrow BaSO_4(s) + 2NaCl(aq)$$

Thermochemistry

In general, chemical reactions either liberate or absorb heat. The energy changes in a reaction are due, to a large extent, to the changes in potential energy that accompany the alterations in the interactions between particles (particularly the breaking and making of chemical bonds). A more fundamental origin of chemical energy lies in the potential and kinetic energy of the atoms, molecules, and subatomic particles that make up a sample, in addition to the situation in which the sample exists—that is, with respect to pressure and volume. Each of these factors contributes to the total capacity these particles have to exchange heat; this property is referred to as **enthalpy**. Enthalpy is symbolized by the letter H. Although it is not possible to measure the enthalpy of a chemical system in an absolute manner, it is possible to measure the *change* in this property as a system changes; the change is consequently symbolized as ΔH. ΔH is often referred to as the heat of reaction. Changes in enthalpy for exothermic and endothermic reactions can be shown graphically, as in Figure 8.1 below.

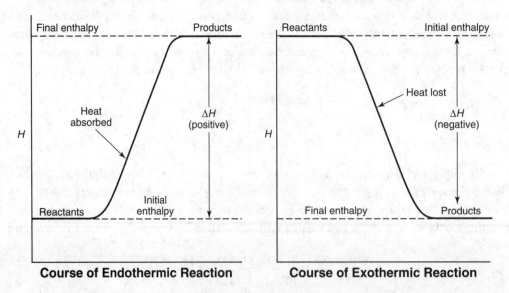

FIGURE 8.1 Changes in Enthalpy for Exothermic and Endothermic Reactions

Notice that the ΔH for the endothermic reaction is positive, while that for the exothermic reaction is negative. This is so because chemists always take the perspective of the chemical system when describing change. In other words, in an exothermic process, the system is releasing energy (i.e., the energy of the system is going down). During an endothermic process, the system is absorbing energy (i.e., the energy of the system is going up). Changes in enthalpy are independent of the path, or series of steps, by which a reaction occurs; this form of change is known as a **state function**. ΔH simply describes the difference in the capacity substances have to exchange heat relative to the initial and final state of the system.

Predicting Reactions

One of the most important topics of chemistry deals with the reasons why reactions take place. Taking each of the previously mentioned reaction types and considering the above-mentioned information about thermochemistry, let's see how a prediction can be made concerning the driving force that makes these reactions occur.

1. Combination (Also Known as Synthesis)

A good source of predictive information (although technically not the only thing to consider) for whether a chemical combination will generate a reaction is a **heat of formation** table. The standard heat of formation, symbolized ΔH_f°, is a particular type of heat of reaction and is associated only with formation reactions. The phase of the compound forming (solid, liquid, or gas) is important to note, as differences seen in the values depend on the compound's state of matter.

If the heat of formation is a relatively large negative value, the combination is likely to occur spontaneously, since one driving force for a reaction is for a chemical system to lower its energy. On the other hand, if the heat of formation is a small negative value or even a positive value, the reaction is likely to be non-spontaneous and also likely to require energy to proceed at any noticeable rate. The examples below are for synthesis reactions that are likely to occur.

TIP

$-\Delta H =$ exothermic reaction

$+\Delta H =$ endothermic reaction

> **EXAMPLE**
>
> $$Zn(s) + S(s) \rightarrow ZnS(s) + 202.7 \text{ kJ} \qquad \Delta H_f = -202.7 \text{ kJ}$$
>
> This means that 1 mole of zinc (65 grams) reacts with 1 mole of sulfur (32 grams) to form 1 mole of zinc sulfide (97 grams) and releases 202.7 kilojoules of heat.

2. Decomposition (Also Known as Analysis)

The prediction of decomposition reactions uses the same source of information, the heat of formation table. If the heat of formation of the reactant is a large exothermic (ΔH is negative) value, the compound will be difficult to decompose since this same quantity of energy must be returned to the compound. A relatively low and negative heat of formation indicates decomposition would not be difficult, such as the decomposition of mercuric oxide with a $\Delta H_f = -90.8$ kJ/mole:

$$2HgO(s) \rightarrow 2Hg(s) + O_2(g) \text{ (Priestley's method of preparation)}$$

A high positive heat of formation indicates extreme instability of a compound, which can explosively decompose.

3. Single Replacement

A prediction of the feasibility of this type of reaction can be based on a comparison of the heat of formation of the original compound and that of the compound to be formed. It is noteworthy that the heats of formation for the elements, in this type of reaction or others, are not included in the analysis. This is so because their ΔH_f° values, by definition, are zero and thus have no impact in the overall energy analysis.

EXAMPLE

$$Zn(s) + 2HCl(aq) \rightarrow ZnCl_2(aq) + H_2(g)$$

Since the value for the heat of formation of HCl is -92.3 kJ per mole and there are two moles of HCl shown as reactants, the value of $2 \times (-92.3$ kJ$)$ would have to be compared to the heat of formation of $ZnCl_2$, which is -415.5 kJ/mol. That comparison shows that the products possess an overall larger negative ΔH value (-415.5 kJ compared to -184.6 kJ). As the reaction is likely to occur with the difference of 230.9 kJ of heat being released, the energy of the chemical system is lowered.

Another way of predicting the spontaneity of single replacement reactions is to check the relative positions of the two elements in the activity series in Table 8.1. If the element that is to replace the other in the compound is higher on the chart, the reaction will occur. If it is below, there will be no reaction.

In predicting the replacement of hydrogen by zinc in hydrochloric acid, as seen in the example above, reference to the activity series shows that zinc will replace hydrogen. Again, predicting this reaction would occur:

$$Zn(s) + 2HCl(aq) \rightarrow ZnCl_2(aq) + H_2(g)$$

In fact, most metals in the activity series would replace hydrogen in an acid solution as hydrogen is fairly low in the activity series. If a metal such as copper were chosen, no reaction would occur, however, as copper is even lower then hydrogen in the series.

$$Cu(s) + HCl(aq) \rightarrow \text{no reaction}$$

TABLE 8.1 Activity Series of Common Elements

		Activity of Metals	Activity of Halogen Nonmetals
Most active	Li		F_2
	Rb	React with cold water and acids, replacing hydrogen. React with oxygen, forming oxides.	Cl_2
	K		Br_2
	Ba		I_2
	Sr		
	Ca		
	Na		
	Mg		
	Al	React with steam (but not cold water) and acids, replacing hydrogen. React with oxygen, forming oxides.	
	Mn		
	Zn		
	Cr		
	Fe		
	Cd		
	Co	Do not react with water. React with acids, replacing hydrogen. React with oxygen, forming oxides.	
	Ni		
	Sn		
	Pb		
	H_2		
	Sb	React with oxygen, forming oxides.	
	Bi		
	Cu		
	Hg		
Least active	Ag	Fairly unreactive, forming oxides only indirectly	
	Pt		
	Au		

4. Double Replacement

For double replacement reactions to go to completion, that is, proceed until the supply of one of the reactants is exhausted, one of the following conditions must be present: (1) an insoluble precipitate is formed, (2) a nonionizing substance is formed, or (3) a gaseous product is given off.

1. To predict the formation of an **insoluble precipitate**, you should have some knowledge of the solubilities of compounds and some general solubility rules.

TABLE 8.2 Solubilities of Compounds

Soluble	Except
Na^+ NH_4^+ } compounds K^+	
Acetates	
Bicarbonates	
Chlorates	
Chlorides...	Ag, Hg, Pb ($PbCl_2$, sol. in hot water)
Nitrates	
Sulfates...	Ba, Ca (slight), Pb
Insoluble	
Carbonates, phosphates..	Na, NH_4, K compounds
Sulfides, hydroxides...	Na, NH_4, K, Ba, Ca

An example of this type of reaction is found when a solution of potassium chloride reacts with a solution of silver nitrate.

$$KCl(aq) + AgNO_3(aq) \rightarrow AgCl(s) + KNO_3(aq)$$

When the reaction is given in its complete ionic form,

$$K^+(aq) + Cl^-(aq) + Ag^+(aq) + NO_3^-(aq) \rightarrow AgCl(s) + K^+(aq) + NO_3^-(aq),$$

it's easy to see that the silver ion combines with the chloride ion to make the insoluble precipitate silver chloride; the potassium and nitrate ions are not really involved in the reaction as they are not changing. They are referred to as *spectator ions*.

2. Another reason for a reaction of this type to go to completion is the formation of a **nonionizing product** such as water. This weak electrolyte keeps its component ions in molecular form and thus eliminates the possibility of reversing the reaction. All neutralization reactions are of this type.

$$H^+(aq) + Cl^-(aq) + Na^+(aq) + OH^-(aq) \rightarrow H_2O(\ell) + Na^+(aq) + Cl^-(aq)$$

This example shows the ions of the reactants, hydrochloric acid and sodium hydroxide, and the nonelectrolyte product water with sodium and chloride ions in solution. Since the water does not ionize to any extent, the reverse reaction cannot occur.

3. The third reason for double displacement to occur is the **evolution of a gaseous product**. An example of this is calcium carbonate reacting with hydrochloric acid:

$$CaCO_3(s) + 2HCl(aq) \rightarrow CaCl_2(aq) + H_2O(\ell) + CO_2(g)$$

Viewed from an exchange-of-partners perspective, the water and the carbon dioxide in the products above are the decomposition products of H_2CO_3, the compound that would have formed along with the $CaCl_2$ when the exchange of partners took place.

Another example of a double replacement reaction that evolves a gaseous product is the reaction of sodium sulfite with an acid:

$$Na_2SO_3(aq) + 2HCl(aq) \rightarrow 2NaCl(aq) + H_2O(\ell) + SO_2(g)$$

In general, acids combining with carbonates or sulfites are good examples of this type of reaction.

Entropy

Predicting the spontaneity of reactions with enthalpy change has been useful in many of the preceding reactions, as lowering the energy of a chemical system is *one* of the fundamental driving forces in nature that causes change. There is, however, another property of a system whose change *also* contributes to the spontaneity of a process. This property, called **entropy**, measures the disorder of a system. In this case, however, an increase in the property promotes change. Unlike enthalpy, the absolute entropy of a system *can* be measured. This is so because a benchmark for complete lack of disorder exists naturally in a perfect crystal at a temperature of 0 Kelvins (absolute zero). The combination of the change in enthalpy and the change in entropy constitutes a complete view in which the spontaneity of a process can be determined. It is because entropy change was not considered in reference to most previous reactions that the word "likely" was used to preface the predictions.

> **TIP**
> Entropy is a measure of the degree of disorder.

Determining Changes in Enthalpy

There are two ways in which chemists routinely determine ΔH values. One is experimentally using calorimetry; the other is theoretically using Hess's Law.

Calorimetry

A **calorimeter** is a laboratory device used to measure the enthalpy change in a chemical or physical process. There are two types of calorimeters, one open, the other closed. In open calorimetry, the chemical system is open to the atmosphere and, as such, is subject to a constant pressure. In this case, the heat that flows into or out of a system is equal to the change in enthalpy. This is not the case in closed calorimetry. A simple open calorimeter consists of an insulating container (often just a Styrofoam cup) housing a known mass of water, along with a thermometer and stirring device. Since a **calorie** is a unit of energy and is defined as the amount of energy needed to raise the temperature of 1 gram of water by 1°C, a calorimeter, as its name implies, *measures energy* change. Based on this definition, the **specific heat** for water—that is, the particular amount of energy needed to raise a particular amount of water by a certain degree—is 1 cal/g°C. Since a joule, the SI unit for energy, is 4.186 times smaller than the calorie, the specific heat of water is also 4.186 J/g°C. Because of the known specific heat of water, the amount of energy change involved in a process can easily be calculated by monitoring the temperature of the water in the calorimeter present to the reaction to see how much it changes during the process. The calculation uses the formula

$$q = mc\Delta T$$

where q denotes the heat flow into or out of the water, m stands for the mass of the water, c represents the specific heat of the water, and ΔT its temperature change.

EXAMPLE

Find the change in enthalpy, ΔH, in kJ/mol for the dissolution of NaOH in water; assume that when 0.10 moles (4.0 g) of the sodium hydroxide is dissolved in 96.0 g of water, the temperature of the water rises from 25.0°C to 34.9°C. Also, assume the specific heat of the overall solution is the same as water and that the overall solution (not just the water) will be what absorbs or releases energy from or to the system. This is typically done in fairly dilute solutions.

Using the equation to find the heat flow *into* the solution (this is the case since the temperature of the solution went up) we find that:

$$q = (100.0 \text{ g})(4.184 \text{ J/g°C})(9.9°C)$$
$$q = 4{,}100 \text{ J}$$

This value represents the energy *gained* by the solution (mostly the water) and therefore the amount of energy *lost* by the chemical system. This value also represents the change in enthalpy for the reaction involving the 4.0 grams of NaOH used (which is only 0.10 moles of NaOH). Recall, however, that heat-of-reaction values are typically reported in units of kJ per mole. Therefore, it would be typical to change the sign on the heat flow, q, from positive to negative and the value from joules to kilojoules. Finally, dividing that amount of energy by 0.10 moles allows for representation of the reaction in a typical manner:

$$NaOH(s) \rightarrow Na^+(aq) + OH^-(aq) + 41 \text{ kJ}$$

or

$$\Delta H = -41 \text{ kJ/mol}$$

The making of this solution would be an exothermic process because the heat of reaction is negative. As 41 kJ of energy would be released if it was to occur, it is therefore shown as a product. The production of this solution would feel warm to the touch.

Heating and Cooling Curves

Besides calorimetry, the equation $q = mc\Delta T$ is used in determining the amount of energy needed to raise the temperature of any substance, not just water in a calorimeter. This is typically the case when evaluating the amount of energy needed to take a substance in the solid state, below its freezing point, to the gaseous state, above its boiling point. Figure 8.2 provides a visual depiction of the process *for water* and is called a **heating curve**, as heat energy is constantly being supplied to the water as time goes by. If viewed in reverse, with energy being removed as time goes by and the gas caused to be turned into a solid, it would be called a **cooling curve**.

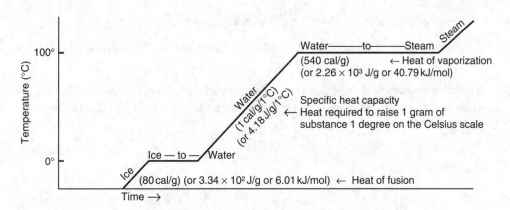

FIGURE 8.2 Changing Ice to Steam

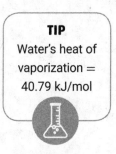

Ice changing to water and then to steam is not a process of continuous and constant change of temperature as time progresses; rather, it is accomplished in stages. Although there are really five stages to analyze when changing a solid below its freezing point to a gas above its boiling point, this graph supplies only numerical information to calculate the amount of energy needed to take ice *at* its freezing point to steam *at* its boiling point. Therefore, we will evaluate only these three energy changes for the system.

Starting at 0°C, water's normal melting point, the ice begins to melt. Notice that the temperature is not increasing despite the constant supply of heat as time passes. The lack of temperature increase indicates that the kinetic energy of the system is not changing; this is so because temperature and kinetic energy are directly related to each other. If the energy is not being used to make the particles move more, what is it doing? The answer is that it is overcoming the attractive forces between the solid particles that hold them in fixed crystalline positions and thus allowing them to exist as non-rigid clusters of particles characteristic of the liquid phase. Consequently, the potential energy of the system must be on the rise, as the energy is being used to overcome the attractive forces between the particles and accomplish the phase change. The energy required to do this, the **heat of fusion**, is an important property of a solid. The heating curve shows various values, utilizing various units, for the heat of fusion of water. Different substances have values different from those shown for water. An upcoming calculation will use the value 6.01 kJ/mol as the heat of fusion of ice. In light of what has been discussed in this chapter, it is simply another particular ΔH value.

Once the entire sample of the solid has melted, the liquid water's temperature begins to rise as more heat is pumped into the system. Since the temperature of the water is rising, the kinetic energy of the system is rising as well. The $q = mc\Delta T$ equation can be used to calculate the energy required to do this. The temperature will continue to rise until it reaches 100.0°C and the liquid water begins to normally boil.

At the normal boiling point, liquid water turns into gaseous water (steam); once again we see a plateau associated with the temperature change. As before, *KE* (kinetic energy) is not increasing, and so *PE* (potential energy) must be going up. This time, the clusters associated with the liquid are being ripped apart into randomness. The energy required to do this, the **heat of vaporization**, is an important property of a liquid. From the heating curve, one value for the heat of vaporization of water is 40.79 kJ/mol.

EXAMPLE

Find the total amount of energy needed to raise a 180 gram sample of ice at 0.0°C to steam at 100.0°C.

This problem involves the three stages of the heating curve described above. First, the ice must be melted. Second, the water must rise in temperature. Third, the water must vaporize to a gas.

Melting

Since the value used for the heat of fusion of ice is in kJ/mol, the amount of ice must be converted to moles. Via mental math, 180 grams converts to 10.0 moles because the molar mass of water is 18.0 grams. Using the heat of fusion, we find that

$$10.0 \text{ moles } (6.01 \text{ kJ/mol}) = 60.1 \text{ kJ (needed to melt the ice)}$$

Raising the Temperature of the Liquid Water

Since the specific heat term has grams in it, the 180-gram value (not moles) should be used for the amount of water in this calculation. ΔT will be 100.0°C as the temperature changes from the normal melting point of water to the normal boiling point of water. Using the q formula, we find that:

$$q = (180. \text{ g})(4.18 \text{ J/g°C})(100.0°C) = 75,200 \text{ J} = 75.2 \text{ kJ}$$

Boiling

Again, moles will be used for the amount of water because the heat of vaporization noted above incorporates that unit. Using heat of vaporization, we find that

$$10.0 \text{ moles } (40.1 \text{ kJ/mol}) = 401 \text{ kJ (needed to boil the liquid)}$$

To find the total energy required, simply add the energy associated with the various stages investigated. The total energy required to change ice at its melting point to steam at its boiling point is 60.1 kJ + 75.2 kJ + 401 kJ or 535 kJ.

Hess's Law of Constant Heat Summation

As suggested previously, there is a theoretical way to determine the heat of reaction for a process. **Hess's Law** takes advantage of the **First Law of Thermodynamics**, which states that like matter, energy is conserved in chemical and physical changes. Consequently, knowledge of ΔH values for certain reactions allows you to find the ΔH value for a reaction whose heat of reaction you don't know. This is so because the change in enthalpy is constant—irrespective of whether a process takes place in one step or many—because energy is conserved. This idea complements previous ideas in this chapter, which stated that heat of reaction is pathway independent (and is called a state function) in that the total enthalpy change is dependent only on the initial and final enthalpies of the reactants and products.

Stated in a useful manner, Hess's Law allows for reactions about which you know ΔH values to be added to find ΔH values for reactions about which you don't know. This is seen in the examples below.

EXAMPLE

Find the heat of reaction, ΔH, for the process

$$C(s) + O_2(g) \rightarrow CO_2(g)$$

when the heats of reaction for the following processes are provided:

$$C(s) + \tfrac{1}{2} O_2(g) \rightarrow CO(g) \qquad \Delta H = -110.5 \text{ kJ}$$

$$CO(g) + \tfrac{1}{2} O_2(g) \rightarrow CO_2(g) \qquad \Delta H = -283.0 \text{ kJ}$$

Addition of the chemical processes above results in the reaction for which the heat of reaction is desired. The CO seen in both the products of the first reaction and the reactants of the second cancel out upon the addition of the reactions. Furthermore, the $\tfrac{1}{2} O_2(g)$ reactant, seen in both reactions, adds to make present $1\ O_2(g)$, as is found in the target reaction (i.e., the reaction for which the heat of reaction is not given). Carbon, C, is found in the reactants upon addition, and CO_2 is likewise found in the products. Hence, the addition of the two "steps" produces the target reaction. As stated before, energy change is constant; it is not contingent upon the number of steps it takes to accomplish a process. Since the two reactions add up to the target reaction, their ΔH values should add up to that of the overall reaction. The ΔH for the target reaction is then -393.5 kJ.

Sometimes reactions with known ΔH values are given as steps for an overall process but cannot be added together directly to achieve the target reaction. These may require some preliminary mathematical manipulation prior to the addition of the steps. The types of mathematical manipulation include:

1. *Flipping a reaction* (i.e., looking at the reaction in reverse). This will cause the substances previously in the reactants to now be in the products and vice versa. As a result, the enthalpy change will become the opposite of what it previously was from a thermodynamic perspective. This is easily shown by changing the sign on ΔH.

2. *Multiplying or dividing a reaction.* Since enthalpy is an extensive property, in which the amount of substance involved in a reaction is important to consider, proper numbers of moles of reactants and products need to be seen in the steps to achieve the target reaction upon addition. Whatever value the reaction is multiplied or divided by should also be applied to the ΔH value.

EXAMPLE

Find the heat of reaction, ΔH, for the process

$$W(s) + C(s) \rightarrow WC(s) \text{ (the target reaction)}$$

with the heats of reaction for the following processes provided:

$$2W(s) + 3O_2(g) \rightarrow 2WO_3(s) \qquad \Delta H = -1{,}680.6 \text{ kJ}$$

$$C(s) + O_2(g) \rightarrow CO_2(g) \qquad \Delta H = -393.5 \text{ kJ}$$

$$2WC(s) + 5O_2(g) \rightarrow 2WO_3(s) + 2CO_2(g) \qquad \Delta H = -2{,}391.6 \text{ kJ}$$

Addition of the reactions above, as written, would not yield the target reaction, and so appropriate mathematical manipulation must be done. Analysis of the reactions shows that the pertinent substance in both the target reaction and the first step is tungsten, W. Since the target reaction indicates only 1 mole of tungsten reacting but the first step indicates 2 moles of tungsten reacting, the first step must be divided by 2 and rewritten as:

$$W(s) + \frac{3}{2}O_2(g) \rightarrow WO_3(s) \qquad \Delta H = -840.3 \text{ kJ}$$

Further analysis of the reactions shows that the pertinent substance in both the target reaction and the second step is carbon, C. Since the target reaction indicates 1 mole of carbon is reacting and the second step indicates the same, no manipulation of step 2 is needed.

$$C(s) + O_2(g) \rightarrow CO_2(g) \qquad \Delta H = -393.5 \text{ kJ}$$

Final analysis of the reactions shows that the pertinent substance in both the target reaction and the third step is tungsten carbide, WC. Since the target reaction indicates 1 mole of tungsten carbide is produced and the third step shows 2 moles of tungsten carbide reacting, the third step must be divided by 2 and flipped.

$$WO_3(s) + CO_2(g) \rightarrow WC(s) + \frac{5}{2}O_2(g) \qquad \Delta H = 1{,}195.8 \text{ kJ}$$

Addition of the three manipulated equations produces the target reaction (after the CO_2, O_2, and WO_3 all cancel) and so the addition of the ΔH values associated with those reactions will give the ΔH for the target reaction.

$$W(s) + C(s) \rightarrow WC(s) \qquad \Delta H = -38 \text{ kJ}$$

Standard heats of formation (introduced at the beginning of this chapter) are values typically found in the appendix of most chemistry textbooks. They serve as a resource to predict the spontaneity of reactions like synthesis, decomposition, and single replacement. They can also be used as a resource to help determine $\Delta H°$ values for reactions as a less cumbersome alternative to the method shown above. Rooted in Hess's Law, the technique is based on the concept that a standard heat of reaction, $\Delta H°$, is equal to the difference between the total enthalpy of the reactants and the total enthalpy of the products. This can be expressed as follows:

$$\Delta H°_{reaction} = \Sigma\, n\Delta H°_{f(products)} - \Sigma\, n\Delta H°_{f(reactants)}$$

where Σ = the summation of

n = the number of moles of the substances involved in the reaction

$\Delta H°_f$ = the standard heat of formation for the substance

EXAMPLE

Find the standard heat of reaction, $\Delta H°_{reaction}$, for the decomposition of sodium chlorate:

$$2NaClO_3(s) \rightarrow 2NaCl(s) + 3O_2(g)$$

$$\Delta H°_f \text{ for } NaClO_3(s) = -358.2 \text{ kJ/mol}$$

$$\Delta H°_f \text{ for } NaCl(s) = -410.5 \text{ kJ/mol}$$

$$\Delta H°_f \text{ for } O_2(g) = 0 \text{ kJ/mol}$$

Using the values above in the equation yields

$$\Delta H^{\circ}_{\text{reaction}} = \Sigma n \Delta H^{\circ}_{f(\text{products})} - \Sigma n \Delta H^{\circ}_{f(\text{reactants})}$$

$$\Delta H^{\circ}_{\text{reaction}} = (2\text{ mol})(-410.5\text{ kJ/mol}) - (2\text{ mol})(-358.2\text{ kJ/mol})$$

$$\Delta H^{\circ}_{\text{reaction}} = -104.6\text{ kJ}$$

Once known, the heat of reaction can be used to determine the energy released or absorbed in a process involving different amounts of substances relative to what is shown in the reaction equation to which the ΔH value is associated.

EXAMPLE

Having found the heat of reaction for the decomposition of sodium chlorate,

$$2\text{NaClO}_3(s) \rightarrow 2\text{NaCl}(s) + 3\text{O}_2(g) \qquad \Delta H^{\circ} = -104.6\text{ kJ},$$

find the total amount of energy released when a 106.5 gram sample of sodium chlorate is decomposed.

Since the reaction equation describes the amount of energy change associated with 2 moles of NaClO_3 decomposing, you must first find the number of moles of NaClO_3 represented by 106.5 grams of the substance. A quick check of the Periodic Table reveals that the molar mass of sodium chlorate is 106.5 grams. Therefore, the amount of NaClO_3 decomposing is simply 1.000 mol. Employing dimensional analysis, complete the calculation using the ΔH value associated with the reaction equation:

$$(1.000\text{ mol NaClO}_3)(-104.6\text{ kJ/2 mol NaClO}_3) = -52.3\text{ kJ}$$

Since the energy change is shown as a negative value, the information should be interpreted to mean that 52.3 kJ of energy will be released upon the decomposition of 106.5 g of sodium chlorate.

CHAPTER 9

Rates of Chemical Reactions

The measurement of reaction rate is based on the rate of appearance of a product or disappearance of a reactant. It is usually expressed in terms of a change in concentration of one of the participants per unit time.

Experiments have shown that for most reactions the concentrations of all participants change most rapidly at the beginning of the reaction; that is, the concentration of the products shows the greatest rate of increase, and the concentrations of the reactants the highest rate of decrease, at this point. This means that the rate of a reaction changes with time. Therefore, a rate must be identified with a specific time.

Factors Affecting Reaction Rates

Five important factors control the rate of a chemical reaction. These are summarized below.

1. *The nature of the reactants.* In chemical reactions, some bonds break and others form. Therefore, the rates of chemical reactions should be affected by the nature of the bonds in the reacting substances.

2. *The surface area exposed.* Since most reactions depend on the reactants coming into contact, the surface exposed proportionally affects the rate of the reaction.

3. *The concentrations.* The reaction rate is usually proportional to the concentrations of the reactants. The usual dependence of the reaction rate on the concentration of the reactants can simply be explained by theorizing that, if more molecules or ions of the reactant are in the reaction area, then there is a greater chance that more reactions will occur. This idea is further developed in the **collision theory** discussed below.

4. **The temperature.** A temperature increase of 10°C above room temperature usually causes the reaction rate to double or triple. The basis for this generality is that, as the temperature increases, the average kinetic energy of the particles involved increases. As a result the particles move faster and have a greater probability of hitting other reactant particles. Because the particles have more energy, they can cause an effective collision, resulting in the chemical reaction that forms the product substance. This is known as collision theory.

5. **The presence of a catalyst.** It is a substance that increases or decreases the rate of a chemical reaction without itself undergoing any permanent chemical change. The catalyst provides an alternative pathway by which the reaction can proceed and in which the activation energy is lower. It thus increases the rate at which the reaction comes to completion or equilibrium. Generally, the term is used for a substance that increases reaction rate (a positive catalyst). Some reactions can be slowed down by negative catalysts.

Activation Energy

TIP

Activation energy is the energy necessary to cause a reaction to occur.

Often a reaction rate may be increased or decreased by affecting the **activation energy**, that is, the energy necessary to cause a reaction to occur. This is shown graphically in Figure 9.1 for the forward reaction.

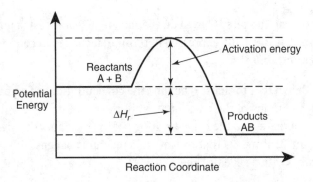

FIGURE 9.1 Potential Energy Diagram for an Exothermic Reaction

A **catalyst**, as explained in the preceding section, is a substance that is introduced into a reaction to speed up the reaction by changing the amount of activation energy needed. The effect of a catalyst used to speed up a reaction can be shown as follows in Figure 9.2:

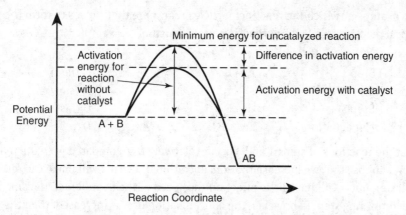

FIGURE 9.2 Potential Energy Diagram for Uncatalyzed/Catalyzed Reaction

> **TIP**
>
> A *catalyst* speeds up the reaction but is not consumed itself.

Reaction Rate Law

The relationship between the rate of a reaction and the masses (expressed as concentrations) of the reacting substances is summarized in the **Law of Mass Action**. It states that the rate of a chemical reaction is proportional to the product of the concentrations of the reactants. For a general reaction between A and B, represented by

$$a\text{A} + b\text{B} \rightarrow \dots$$

the rate law expression is

$$r \propto [\text{A}]^a[\text{B}]^b$$

or, inserting a constant of proportionality that mathematically changes the expression to an equality, we have:

$$r = k[\text{A}]^a[\text{B}]^b$$

Here k is called the **specific rate** constant for the reaction at the temperature of the reaction.

The exponents a and b may be added to give the total reaction order. For example:

$$\text{H}_2(g) + \text{I}_2(g) \rightarrow 2\text{HI}(g)$$
$$r = k[\text{H}_2]^1[\text{I}_2]^1$$

The sum of the exponents is $1 + 1 = 2$, and therefore we have a second-order reaction.

Reaction Mechanism and Rates of Reaction

The beginning of this chapter stated that the reaction rate is *usually* proportional to the concentrations of the reactants. This occurs because some reactions do not directly occur between the reactants but may go through intermediate steps to get to the final product.

The series of steps by which the reacting particles rearrange themselves to form the products of a chemical reaction is called the **reaction mechanism**. For example:

Step 1	$A + B$	$\rightarrow$	I_1 (fast)
Step 2	$A + I_1$	$\rightarrow$	I_2 (slow)
Step 3	$C + I_2$	$\rightarrow$	D (fast)
Net equation	$2A + B + C$	$\rightarrow$	D

Notice that the reactions of steps 1 and 3 occur relatively fast compared with the reaction of step 2. Now suppose that we increase the concentration of C. This will make the reaction of step 3 go faster, but it will have little effect on the speed of the overall reaction since step 2 is the rate-determining step. If, however, the concentration of A is increased, then the overall reaction rate will increase because step 2 will be accelerated. Knowing the reaction mechanism provides the basis for predicting the effect of a concentration change of a reactant on the overall rate of reaction. Another way of determining the effect of concentration changes is actual experimentation.

CHAPTER 10

Chemical Equilibrium

Learning Objectives

In this chapter, you will learn how to:

○ Explain the development of an equilibrium condition and how it is expressed as an equilibrium constant, and use it mathematically.

○ Describe Le Châtelier's Principle and how changes in temperature, pressure, and concentrations affect an equilibrium.

○ Solve problems dealing with ionization of water, solubility products, and the common ion effect.

○ Explain the relationship of enthalpy and entropy as driving forces in a reaction and how they are combined in the Gibbs equation.

In some reactions no product is formed to allow the reaction to go to completion; that is, the reactants and products can still interact in both directions. This can be shown as follows:

$$A + B \rightleftharpoons C + D$$

The double arrow indicates that C and D can react to form A and B, while A and B react to form C and D.

Reversible Reactions and Equilibrium

The reaction is said to have reached **equilibrium** when the forward reaction rate is equal to the reverse reaction rate. Notice that this is a dynamic condition, NOT a static one, although in appearance the reaction *seems* to have stopped. An example of an equilibrium is a crystal of copper sulfate in a saturated solution of copper sulfate. Although to the observer the crystal seems to remain unchanged, there is actually an equal exchange of crystal material with the copper sulfate in solution. As some solute comes out of solution, an equal amount is going into solution.

To express the rate of reaction in numerical terms, we can use the **Law of Mass Action**, which states: The rate of a chemical reaction is proportional to the product of the concentrations of the reacting substances. The concentrations are expressed in moles of gas per liter of volume or moles of solute per liter of solution. Suppose, for example, that 1 mole/liter of gas A_2 (diatomic molecule) is allowed to react with 1 mole/liter of another diatomic gas, B_2, and they form gas AB; let R be the rate for the forward reaction forming AB. The bracketed

> **TIP**
> Equilibrium is reached when the forward and reverse reaction rate are equal.

symbols $[A_2]$ and $[B_2]$ represent the concentrations in moles per liter for these diatomic molecules. Then $A_2 + B_2 \rightarrow 2AB$ has the rate expression

$$R \propto [A_2] \times [B_2]$$

where $\propto$ is the symbol for "proportional to." When $[A_2]$ and $[B_2]$ are both 1 mole/liter, the reaction rate is a certain constant value (k_1) at a fixed temperature.

$$R = k_1 \ (k_1 \text{ is called the rate constant})$$

For any concentrations of A and B, the reaction rate is:

$$R = k_1 \times [A_2] \times [B_2]$$

If $[A_2]$ is 3 moles/liter and $[B_2]$ is 2 moles/liter, the equation becomes:

$$R = k_1 \times 3 \times 2 = 6k_1$$

The reaction rate is six times the value for a 1 mole/liter concentration of each reactant.

At the fixed temperature of the forward reaction, AB molecules are also decomposing. If we designate this reverse reaction as R', then, since

$$2AB \text{ (or } AB + AB) \rightarrow A_2 + B_2$$

two molecules of AB must decompose to form one molecule of A_2 and one of B_2. Thus, the reverse reaction in this equation is proportional to the square of the molecular concentration of AB:

$$R' \propto [AB] \times [AB]$$
$$\text{or } R' \propto [AB]^2$$
$$\text{and } R' \propto k_2 \times [AB]^2$$

where k_2 represents the rate of decomposition of AB at the fixed temperature. Both reactions can be shown in this manner:

$$A_2 + B_2 \rightleftharpoons 2AB \text{ (note double arrow)}$$

When the first reaction begins to produce AB, some AB is available for the reverse reaction. If the initial condition is only the presence of A_2 and B_2 gases, then the forward reaction will occur rapidly to produce AB. As the concentration of AB increases, the reverse reaction will increase. At the same time, the concentrations of A_2 and B_2 will be decreasing and consequently the forward reaction rate will decrease. Eventually the two rates will be equal, that is, $R = R'$. At this point, equilibrium has been established, and:

$$k_1[A_2] \times [B_2] = k_2[AB]^2$$

or

$$\frac{k_1}{k_2} = \frac{[AB]^2}{[A_2] \times [B_2]} = K$$

The convention is that k_1 (forward reaction) is placed over k_2 (reverse reaction) to get this expression. Then k_1/k_2 can be replaced by K_{eq}, which is called the **equilibrium constant** for this reaction under the particular conditions.

In another general example:

$$a\text{A} + b\text{B} \rightleftharpoons c\text{C} + d\text{D}$$

the reaction rates can be expressed as

$$R = k_1[\text{A}]^a \times [\text{B}]^b$$
$$R' = k_2[\text{C}]^c \times [\text{D}]^d$$

Note that the values of k_1 and k_2 are different, but that each is a constant for the conditions of the reaction. At the start of the reaction, [A] and [B] will be at their greatest values, and R will be large; [C], [D], and R' will be zero. Gradually R will decrease and R' will become equal. At this point, the reverse reaction is forming the original reactants just as rapidly as they are being used up by the forward reaction. Therefore, no further change in R, R', or any of the concentrations will occur.

If we set R' equal to R, we have:

$$k_2 \times [\text{C}]^c \times [\text{D}]^d = k_1 \times [\text{A}]^a \times [\text{B}]^b$$

or

$$\frac{[\text{C}]^c \times [\text{D}]^d}{[\text{A}]^a \times [\text{B}]^b} = \frac{k_1}{k_2} = K$$

This process of two substances, A and B, reacting to form products C and D and the reverse can be shown graphically to represent what happens as equilibrium is established. The hypothetical equilibrium reaction is described by the following general equation:

$$a\text{A} + b\text{B} \rightleftharpoons c\text{C} + d\text{D}$$

At the beginning (time t_0), the concentrations of C and D are zero and those of A and B are maximum. The graph below shows that over time the rate of the forward reaction decreases as A and B are used up. Meanwhile, the rate of the reverse reaction increases as C and D are formed. When these two reaction rates become equal (at time t_1), equilibrium is established. The individual concentrations of A, B, C, and D no longer change if conditions remain the same. To an observer, it appears that all reaction has stopped; in fact, however, both reactions are occurring at the same rate seen graphically in Figure 10.1.

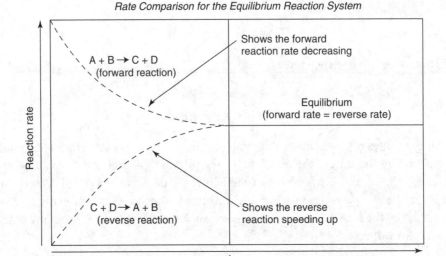

FIGURE 10.1 Change in Reaction Rate as Equilibrium Is Reached

We see that, for the given reaction and the given conditions, K_{eq} is a constant, called the equilibrium constant. If K_{eq} is large, this means that equilibrium does not occur until the concentrations of the original reactants are small and those of the products large. A small value of K_{eq} means that equilibrium occurs almost at once and relatively little product is produced.

The equilibrium constant, K_{eq}, has been determined experimentally for many reactions, and the values are given in chemical handbooks.

Suppose we find K_{eq} for reacting H_2 and I_2 at 490°C to be equal to 45.9. Then the equilibrium constant for this reaction:

$$H_2(g) + I_2(g) \rightleftharpoons 2HI(g) \text{ at } 490°C$$

is

$$K = \frac{[HI]^2}{[H_2][I_2]} = 45.9$$

EXAMPLE

Find the concentrations at equilibrium for the same conditions as in the preceding example except that only 2.00 moles of HI are injected into the box.

$$[H_2] = 0 \text{ mol/L}$$
$$[I_2] = 0 \text{ mol/L}$$
$$[HI] = 2.00 \text{ mol/L}$$

At equilibrium:

$$[HI] = (2.00 - x) \text{ mol/L}$$

Le Châtelier's Principle

A general law, **Le Châtelier's Principle** can be used to explain the results of applying any change of condition (stress) on a system in equilibrium. *It states that if a stress is placed upon a system in equilibrium, the equilibrium is displaced in the direction that counteracts the effect of the stress.* An increase in concentration of a substance favors the reaction that uses up that substance and lowers its concentration. A rise in temperature favors the reaction that absorbs heat and so tends to lower the temperature. These ideas are further developed below.

Effect of Changing the Concentrations

When a system at equilibrium is disturbed by adding or removing one of the substances (thus changing its concentration), all the concentrations will change until a new equilibrium point is reached with the same value of K_{eq}.

If the concentration of a reactant in the forward action is increased, the equilibrium is displaced to the right, favoring the forward reaction. If the concentration of a reactant in the reverse reaction is increased, the equilibrium is displaced to the left. Decreases in concentration will produce effects opposite to those produced by increases.

> **TIP**
> At equilibrium, K_{eq} stays the same at a given temperature.

Effect of Temperature on Equilibrium

If the temperature of a given equilibrium reaction is changed, the reaction will shift to a new equilibrium point. If the temperature of a system in equilibrium is raised, the equilibrium is shifted in the direction that absorbs heat. Note that the shift in equilibrium as a result of temperature change is actually a change in the value of the equilibrium constant. This is different from the effect of changing the concentration of a reactant; when concentrations are changed, the equilibrium shifts to a condition that maintains the same equilibrium constant.

Effect of Pressure on Equilibrium

A change in pressure affects only equilibria in which a gas or gases are reactants or products. Le Châtelier's Law can be used to predict the direction of displacement. If it is assumed that if the total space in which the reaction occurs is constant, the pressure will depend on the total number of molecules in that space. An increase in the number of molecules will increase pressure; a decrease in the number of molecules will decrease pressure. If the pressure is increased, the reaction that will be favored is the one that will lower the pressure, that is, decrease the number of molecules.

An example of the application of these principles is the Haber process of making ammonia. The reaction is:

$$N_2(g) + 3H_2(g) \rightleftharpoons 2NH_3(g) + heat \text{ (at equilibrium)}$$

If the concentrations of the nitrogen and hydrogen are increased, the forward reaction is increased. At the same time, if the ammonia produced is removed by dissolving it into water, the forward reaction is again favored.

Because the reaction is exothermic, the addition of heat must be considered with care. Increasing the temperature causes an increase in molecular motion and collisions, thus allowing the product to form more readily. At the same time, the equilibrium equation shows that the reverse reaction is favored by the increased temperature, so a compromise temperature of about 500°C is used to get the best yield.

An increase in pressure will cause the forward reaction to be favored since the equation shows that four molecules of reactants are forming two molecules of products. This effect tends to reduce the increase in pressure by the formation of more ammonia.

Equilibria in Heterogeneous Systems

The examples so far have involved systems made up of only gaseous substances. Expression of the K values of systems changes when other phases are present.

Equilibrium Constant for Systems Involving Solids

If the experimental data for this reaction are studied:

$$CaCO_3(s) \rightleftharpoons CaO(s) + CO_2(g)$$

it is found that at a given temperature an equilibrium is established in which the concentration of CO_2 is constant. It is also true that the concentrations of the solids have no effect on the CO_2 concentration as long as both solids are present. Therefore, the K_{eq}, which would conventionally be written like this:

$$K = \frac{[CaO][CO_2]}{[CaCO_3]}$$

can be modified by incorporating the concentrations of the two solids. This can be done since the concentration of solids is fixed. It becomes a new constant K, known as:

$$K = [CO_2]$$

Any heterogeneous reaction involving gases does not include the concentrations of pure solids. As another example, K for the reaction:

$$NH_4Cl(s) \rightleftharpoons NH_3(g) + HCl(g)$$

is

$$K = [NH_3][HCl]$$

Acid Ionization Constants

When a weak acid does not ionize completely in a solution, an equilibrium is reached between the acid molecule and its ions. The mass action expression can be used to derive an equilibrium constant, called the **acid dissociation constant**, for this condition. For example, an acetic acid solution ionizing is shown as:

$$HC_2H_3O_2(aq) + H_2O(l) \rightleftharpoons H_3O^+(aq) + C_2H_3O_2^-(aq)$$

$$K = \frac{[H_3O^+][C_2H_3O_2^-]}{[HC_2H_3O_2][H_2O]}$$

The concentration of water in moles/liter is found by dividing the mass of 1 liter of water (which is 1,000 g at 4°C) by its gram-molecular mass, 18 grams, giving H_2O a value of 55.6 moles/liter. Because this number is so large compared with the other numbers involved in the equilibrium constant, it is practically constant and is incorporated into a new equilibrium constant, designated as K_a. The new expression is:

$$K_a = \frac{[H_3O^+][C_2H_3O_2^-]}{[HC_2H_3O_2]}$$

TIP

K_a incorporates the concentration of water.

Ionization constants have been found experimentally for many substances and are listed in chemical tables. The ionization constants of ammonia and acetic acid are about 1.8×10^{-5}. For boric acid $K_a = 5.8 \times 10^{-10}$, and for carbonic acid $K_a = 4.3 \times 10^{-7}$.

If the concentrations of the ions present in the solution of a weak electrolyte are known, the value of the ionization constant can be calculated. Also, if the value of K_a is known, the concentrations of the ions can be calculated.

A small value for K_a means that the concentration of the un-ionized molecule must be relatively large compared with the ion concentrations. Conversely, a large value for K_a means that the concentrations of ions are relatively high. Therefore, the smaller the ionization constant of an acid, the weaker the acid. Thus, of the three acids referred to above, the ionization constants show that the weakest is boric acid, and the strongest, acetic acid. It should be remembered that, in all cases where ionization constants are used, the electrolytes must be weak in order to be involved in ionic equilibria.

Ionization Constant of Water

Because water is a very weak electrolyte, its ionization constant can be expressed as follows:

$$2H_2O(l) \rightleftharpoons H_3O^+(aq) + OH^-(aq)$$

(Equilibrium constant) $K = \dfrac{[H_3O^+][OH^-]}{[H_2O]^2}$ (since $[H_2O^2]$ remains relatively constant, it is incorporated into K_w)

(Dissociation constant) $K_w = [H_3O^+][OH^-] = 1 \times 10^{-14}$ at 25°C

TIP

K_w incorporates the $[H_2O]^2$.

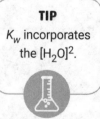

Solubility Products

A saturated solution of a substance has been defined as an equilibrium condition between the solute and its ions. For example:

$$AgCl(s) \rightleftharpoons Ag^+(aq) + Cl^-(aq)$$

The equilibrium constant would be:

$$K = \frac{[Ag^+][Cl^-]}{[AgCl]}$$

Since the concentration of the solute remains constant for that temperature, the [AgCl] is incorporated into the K to give the K_{sp}, called the **solubility constant**:

$$K_{sp} = [Ag^+][Cl^-] = 1.2 \times 10^{-10} \text{ at } 25°C$$

TIP

K_{sp} incorporates the concentration of the solute.

This setup can be used to solve problems in which the ionic concentrations are given and the K_{sp} is to be found or the K_{sp} is given and the ionic concentrations are to be determined.

EXAMPLE

Predicting the formation of a precipitate.

In some cases, the solubility products of solutions can be used to predict the formation of a precipitate.

Suppose we have two solutions. One solution contains 1.00×10^{-3} mole of silver nitrate, $AgNO_3$, per liter. The other solution contains 1.00×10^{-2} mole of sodium chloride, NaCl, per liter. If 1 liter of the $AgNO_3$ solution and 1 liter of the NaCl solution are mixed to make a 2-liter mixture, will a precipitate of AgCl form?

In the $AgNO_3$ solution, the concentrations are:

$$[Ag^+] = 1.00 \times 10^{-3} \text{ mol/L and } [NO_3^-] = 1.00 \times 10^{-3} \text{ mol/L}$$

In the NaCl solution, the concentrations are:

$$[Na^+] = 1.00 \times 10^{-2} \text{ mol/L and } [Cl^-] = 1.00 \times 10^{-2} \text{ mol/L}$$

When 1 liter of one of these solutions is mixed with 1 liter of the other solution to form a total volume of 2 liters, the concentrations will be halved.

In the mixture then, the initial concentrations will be:

$$[Ag^+] = 0.50 \times 10^{-3} \quad \text{or} \quad 5.0 \times 10^{-4} \text{ mol/L}$$
$$[Cl^-] = 0.50 \times 10^{-2} \quad \text{or} \quad 5.0 \times 10^{-3} \text{ mol/L}$$

For the K_{sp} of AgCl:

$$[Ag^+][Cl^-] = [5.0 \times 10^{-4}][5.0 \times 10^{-3}]$$
$$[Ag^+][Cl^-] = 25 \times 10^{-7} \quad \text{or} \quad \underline{2.5 \times 10^{-6}}$$

This is far greater than 1.7×10^{-10}, which is the K_{sp} of AgCl. These concentrations cannot exist, and Ag^+ and Cl^- will combine to form solid AgCl precipitate. Only enough Ag^+ ions and Cl^- ions will remain to make the product of the respective ion concentrations equal 1.7×10^{-10}.

Common Ion Effect

When a reaction has reached equilibrium, and an outside source adds more of one of the ions that is already in solution, the result is to cause the reverse reaction to occur at a faster rate and reestablish the equilibrium. This is called the **common ion effect**. For example, in this equilibrium reaction:

$$NaCl(s) \rightleftharpoons Na^+(aq) + Cl^-(aq)$$

the addition of concentrated HCl (12M) adds H^+ and Cl^- both at a concentration of 12 M. This increases the concentration of the Cl^- and disturbs the equilibrium. The reaction will shift to the left and cause some solid NaCl to come out of solution.

The "common" ion is the one already present in an equilibrium before a substance is added that increases the concentration of that ion. The effect is to reverse the solution reaction and to decrease the solubility of the original substance, as shown in the above example.

Driving Forces of Reactions

Some reactions are said to go to completion because the equilibrium condition is achieved when practically all the reactants have been converted to products. At the other extreme, some reactions reach equilibrium immediately with very little product being formed. These two examples are representative of very large K values and very small K values, respectively. There are essentially two driving forces that control the extent of a reaction and determine when equilibrium will be established. These are the drive to the lowest heat content, or **enthalpy**, and the drive to the greatest randomness or disorder, which is called **entropy**. Reactions with negative ΔH's (enthalpy or heat content) are exothermic, and reactions with positive ΔS's (entropy or randomness) are proceeding to greater randomness.

The **Second Law of Thermodynamics** states that the entropy of the universe increases for any spontaneous process. This means that the entropy of a system may increase or decrease but that, if it decreases, then the entropy of the surroundings must increase to a greater extent so that the overall change in the universe is positive. In other words:

$$\Delta S_{universe} = \Delta S_{system} + \Delta S_{surroundings}$$

The following is a list of conditions in which ΔS is positive for the system:

1. When a gas is formed from a solid, for example,
$$CaCO_3(s) \rightarrow CaO(s) + CO_2(g).$$

2. When a gas is evolved from a solution, for example,
$$Zn(s) + 2H^+(aq) \rightarrow H_2(g) + Zn^{2+}(aq).$$

3. When the number of moles of gaseous product exceeds the moles of gaseous reactant, for example,
$$2C_2H_6(g) + 7O_2(g) \rightarrow 4CO_2(g) + 6H_2O(g).$$

4. When crystals dissolve in water, for example,
$$NaCl(s) \rightarrow Na^+(aq) + Cl^-(aq).$$

> **TIP**
> When the ΔS is positive for the system, it means greater disorder.

Looking at specific examples, we find that in some cases endothermic reactions occur when the products provide greater randomness or positive entropy. This reaction is an example:

$$CaCO_3(s) \rightleftharpoons CaO(s) + CO_2(g)$$

The production of the gas and thus greater entropy might be expected to take this reaction almost to completion. However, this does not occur because another force is hampering this reaction. It is the absorption of energy, and thus the increase in enthalpy, as the $CaCO_3$ is heated.

The equilibrium condition, then, at a particular temperature, is a compromise between the increase in entropy and the increase in enthalpy of the system.

The Haber process of making ammonia is another example of this compromise of driving forces that affect the establishment of an equilibrium. In this reaction

$$N_2(g) + 3H_2(g) \rightleftharpoons 2NH_3(g) + heat$$

the forward reaction to reach the lowest heat content and thus release energy cannot go to completion because the force to maximum randomness is driving the reverse reaction.

Change in Free Energy of a System—the Gibbs Equation

These factors, enthalpy and entropy, can be combined in an equation that summarizes the change of **free energy** in a system. This is designated as ΔG. The relationship is

$$\Delta G = \Delta H - T\Delta S \quad (T \text{ is temperature in kelvins})$$

and is called the **Gibbs free-energy equation**.

The sign of ΔG can be used to predict the spontaneity of a reaction at constant temperature and pressure. If ΔG is negative, the reaction is (probably) spontaneous; if ΔG is positive, the reaction is improbable; and if ΔG is 0, the system is at equilibrium and there is no net reaction.

The ways in which the factors in the equation affect ΔG are shown in Table 10.1:

TIP

Free energy, ΔG, depends on ΔH (enthalpy) and ΔS (entropy).

TABLE 10.1 Spontaneity of a Reaction Considering Change in Free Energy

ΔH	ΔS	ΔG	Will It Happen?	Comment
Exothermic ($-$)	$+$	Always $-$	Yes	No exceptions
Exothermic ($-$)	$-$	$-$ At lower temperatures	Probably	At low temperature
Endothermic ($+$)	$+$	$-$ At higher temperatures	Probably	At high temperature
Endothermic ($+$)	$-$	Never $-$	No	No exceptions

This drive to achieve a minimum of free energy may be interpreted as the driving force of a chemical reaction.

Acids, Bases, and Salts

Learning Objectives

In this chapter, you will learn how to:

- Describe the properties of an Arrhenius acid and base, and know the name, formula, and degree of ionization of common acids and bases.
- Explain the Brønsted-Lowry Theory of acids and conjugate bases.
- Determine pH and pOH of solutions.
- Solve titration problems and the use of indicators in the process.
- Describe how a buffer works.
- Explain the formation and naming of salts.
- Explain amphoteric substances in relation to acid-base theory.

Definitions and Properties

What defines an acid? A base? A salt? You must know the different ways that these interact. You must also know how to determine the pH of a substance and how to neutralize that same substance.

Acids

There are some characteristic properties by which an **acid** may be defined. The most important are:

1. *Water (aqueous) solutions of acids conduct electricity.* The degree of conduction depends on the acid's degree of ionization. A few acids ionize almost completely, while others ionize to only a slight degree.

2. *Acids will react with metals that are more active than hydrogen ions to liberate hydrogen.* (Some acids are also strong oxidizing agents and will not release hydrogen. Somewhat concentrated nitric acid is such an acid.)

3. *Acids have the ability to change the color of indicators.* Some common indicators are litmus and phenolphthalein. **Litmus** is a dyestuff obtained from plant life. When litmus is added to an acidic solution, or paper impregnated with litmus is dipped into an acid, the neutral purple color changes to pink-red. **Phenolphthalein** is pink in a basic solution and becomes colorless in a neutral or acid solution.

Learn the names and formulas of these common acids.

TABLE 11.1 Degrees of Ionization of Common Acids

Completely or Nearly Completely Ionized	Moderately Ionized	Slightly Ionized
Nitric HNO_3	Oxalic $H_2C_2O_4$	Hydrofluoric HF
Hydrochloric HCl	Phosphoric H_3PO_4	Acetic $HC_2H_3O_2$
Sulfuric H_2SO_4	Sulfurous H_2SO_3	Carbonic H_2CO_3
Hydriodic HI		Hydrosulfuric H_2S
Hydrobromic HBr		(Most others)

4. *Acids react with bases so that the properties of both are lost to form water and a salt.* This is called **neutralization**. The general equation is:

$$\text{Acid} + \text{Base} \rightarrow \text{Salt} + \text{Water}$$

An example is:

$$Mg(OH)_2(aq) + H_2SO_4(aq) \rightarrow MgSO_4(aq) + 2H_2O(\ell)$$

5. *Acids react with carbonates to release carbon dioxide.* An example:

$$CaCO_3(s) + 2HCl(aq) \rightarrow CaCl_2(aq) + H_2CO_3 \text{ (unstable and decomposes)}$$
$$\rightarrow H_2O(\ell) + CO_2(g)$$

The most common theory used in first-year chemistry is the **Arrhenius Theory**, which states that an acid is a substance that yields hydrogen ions in an aqueous solution. Although we speak of these hydrogen ions in the solution, they are really not separate ions but become attached to the oxygen of the polar water molecule to form the H_3O^+ ion (the hydronium ion). Thus, it is really this hydronium ion we are concerned with in an acid solution.

The general reaction for the dissociation of an acid, HX, is commonly written as

$$HX \rightleftharpoons H^+ + X^-$$

To show the formation of the hydronium ion, H_3O+, the complete equation is:

$$HX + H_2O \rightleftharpoons H_3O^+ + X^-$$

Bases

Bases may also be defined by some operational definitions that are based on experimental observations. Some of the important ones are as follows:

1. *Bases are conductors of electricity in an aqueous solution.* Their degrees of conduction depend on their degrees of ionization. The degrees of ionization of some common bases are shown in Table 11.2.

TABLE 11.2 Degrees of Ionization of Common Bases

Completely or Nearly Completely Ionized	Slightly Ionized
Potassium hydroxide KOH	Ammonium hydroxide $NH_4(OH)$
Sodium hydroxide NaOH	(All others)
Barium hydroxide $Ba(OH)_2$	
Strontium hydroxide $Sr(OH)_2$	
Calcium hydroxide $Ca(OH)_2$	

TIP
Learn the names and formulas of these common bases.

2. *Bases cause a color change in indicators.* Litmus changes from red to blue in a basic solution, and phenolphthalein turns pink from its colorless state.

3. *Bases react with acids to neutralize each other and form a salt and water.*

4. *Bases react with fats to form a class of compounds called soaps.* Earlier generations used this method to make their own soap.

5. *Aqueous solutions of bases feel slippery, and the stronger bases are very caustic to the skin.*

The Arrhenius Theory defines a base as a substance that yields hydroxide ions (OH^-) in an aqueous solution.

Some common bases have familiar names, for example:

Sodium hydroxide   = lye, caustic soda

Potassium hydroxide   = caustic potash
Calcium hydroxide   = slaked lime, hydrated lime, limewater
Ammonium hydroxide   = ammonia water, household ammonia

Much of the sodium hydroxide produced today comes from the Hooker cell electrolysis apparatus. When an electric current is passed through a saltwater solution, hydrogen, chlorine, and sodium hydroxide are the products. The formula for this equation is:

$$2NaCl(aq) + 2HOH(\ell) \xrightarrow{\text{electrical energy}} H_2(g) + Cl_2(g) + 2NaOH(aq)$$

Broader Acid-Base Theories

Besides the common Arrhenius Theory of acids and bases discussed for aqueous solutions, two other theories, the Brønsted-Lowry Theory and the Lewis Theory, are widely used.

The Brønsted-Lowry Theory (1923) defines acids as proton donors and bases as proton acceptors. This definition agrees with the aqueous solution definition of an acid giving up hydrogen ions in solution, but goes beyond to other cases as well.

An example is the reaction of dry HCl gas with ammonia gas to form the white solid NH_4Cl.

$$HCl(g) + NH_3(g) \rightarrow NH_4Cl(s)$$

The HCl is the proton donor or acid, and the ammonia is a Brønsted-Lowry base that accepts the proton.

Conjugate Acids and Bases

In an acid-base reaction, the original acid gives up its proton to become a **conjugate base**. In other words, after losing its proton, the remaining ion is capable of gaining a proton, thus qualifying as a base. The original base accepts a proton, so it now is classified as a **conjugate acid** since it can release this newly acquired proton and thus behave like an acid.

Some examples are given in Figure 11.1 below:

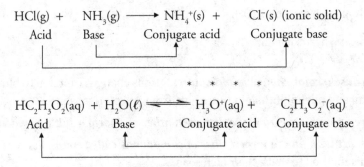

FIGURE 11.1 Examples of Conjugate Acids and Bases

Strength of Conjugate Acids and Bases

The extent of the reaction between a Brønsted-Lowry acid and base depends on the relative strengths of the acids and bases involved. Consider the following example. Hydrochloric is a strong acid. It gives up protons readily. It follows that the Cl^- ion has little tendency to attract and retain a proton. Consequently, the Cl^- ion is an extremely weak base.

$$\underset{\text{strong acid}}{HCl(g)} + \underset{\text{base}}{H_2O(\ell)} \rightarrow \underset{\text{acid}}{H_3O^+(aq)} + \underset{\text{weak base}}{Cl^-(aq)}$$

This observation leads to an important conclusion: the stronger an acid is, the weaker its conjugate base; the stronger a base is, the weaker its conjugate acid. This concept allows strengths of different acids and bases to be compared to predict the outcome of a reaction. As an example, consider the reaction of perchloric acid, $HClO_4$, and water.

$$\underset{\text{stronger acid}}{HClO_4(aq)} + \underset{\text{stronger base}}{H_2O(\ell)} \rightarrow \underset{\text{weaker acid}}{H_3O^+(aq)} + \underset{\text{weaker base}}{ClO_4^-(aq)}$$

Another important general conclusion is that proton-transfer reactions favor the production of the weaker acid and the weaker base. For a reaction to approach completion, the reactants must be much stronger as an acid and as a base than the products.

The Lewis Theory (1916) defines acids and bases in terms of the electron-pair concept, which is probably the most generally useful concept of acids and bases. According to the Lewis definition, an acid is an electron-pair acceptor, and a base is an electron-pair donor. An example seen in Figure 11.2 below is the formation of ammonium ions from ammonia gas and hydrogen ions.

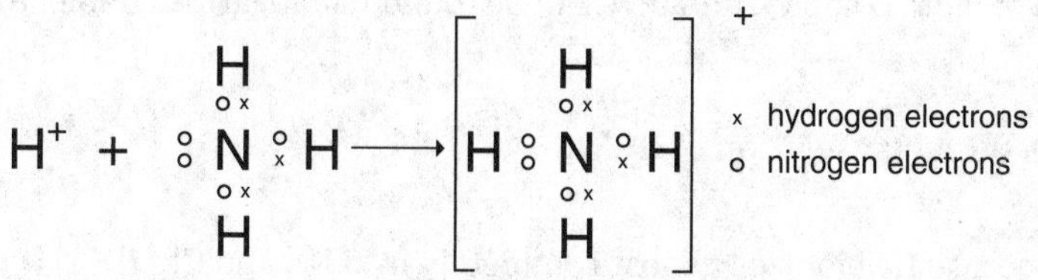

FIGURE 11.2 Formation of Ammonium Ions

x hydrogen electrons
o nitrogen electrons

Notice that the hydrogen ion is in fact accepting the electron pair of the ammonia, so it is a Lewis acid. The ammonia is donating its electron pair, so it is a Lewis base.

Another example is boron trifluoride. It is an excellent Lewis acid. It forms a fourth covalent bond with many molecules and ions. Its reaction with a fluoride ion is shown below in Figure 11.3.

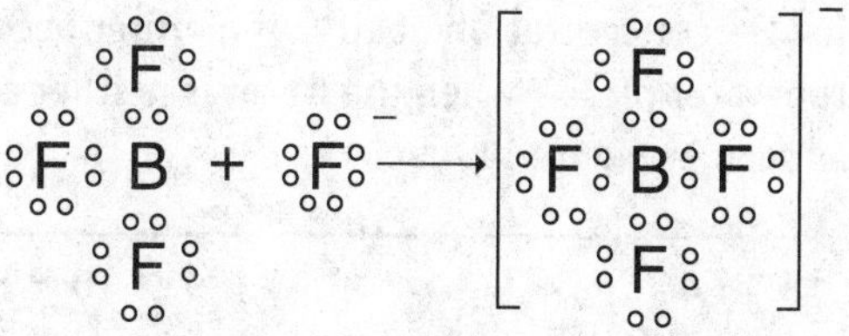

FIGURE 11.3 Boron Trifluoride Reaction with Fluoride Ion

$$BF_3(aq) + F^-(aq) \rightarrow BF_4^-(aq)$$

The acid-base systems are summarized below in Table 11.3.

TABLE 11.3 Acid-Base Systems

Type	Acid	Base
Arrhenius	H^+ or H_3O^+ producer	OH^- producer
Brønsted-Lowry	proton (H^+) donor	proton (H^+) acceptor
Lewis	electron-pair acceptor	electron-pair donor

Acid Concentration Expressed as pH

Frequently, acid and base concentrations are expressed by means of the **pH** system. The pH can be defined as $-\log [H^+]$, where $[H^+]$ is the concentration of hydrogen ions expressed in moles per liter. The logarithm is the exponent of 10 when the number is written in the base 10. For example:

$$100 = 10^2 \text{ so logarithm of 100, base 10} = 2$$
$$10,000 = 10^4 \text{ so logarithm of 10,000, base 10} = 4$$
$$0.01 = 10^{-2} \text{ so logarithm of 0.01, base 10} = -2$$

REMEMBER
pH = −log[H⁺]

The logarithms of more complex numbers can be found in a logarithm table. An example of a pH problem is:

Find the pH of a 0.1 molar solution of HCl.

1st step.	Because HCl ionizes almost completely into H^+ and Cl^-, $[H^+] = 0.1$ mole/liter.
2nd step.	By definition
	$pH = -\log [H^+]$
so	
	$pH = -\log [10^{-1}]$
3rd step.	The logarithm of 10^{-1} is -1
so	
	$pH = -(-1)$
4th step.	The pH then is $= 1$.

Because water has a normal H^+ concentration of 10^{-7} mole/liter because of the slight ionization of water molecules, the water pH is 7 when the water is neither acid nor base. The normal pH range is from 0 to 14 as seen in Figure 11.4 below.

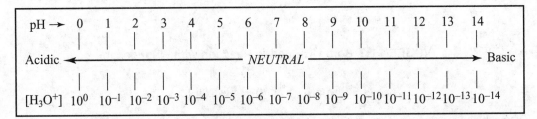

FIGURE 11.4 pH Range

The pOH is the negative logarithm of the hydroxide ion concentration:

$$pOH = -\log [OH^-]$$

If the concentration of the hydroxide ion is 10^{-9} M, then the pOH of the solution is $+9$.

From the equation

$$[H^+][OH^-] = 1.0 \times 10^{-14} \text{ at } 298 \text{ K}$$

the following relationship can be derived:

$$pH + pOH = 14.00$$

In other words, the sum of the pH and pOH of an aqueous solution at 298 K must always equal 14.00. For example, if the pOH of a solution is 9.00, then its pH must be 5.00.

EXAMPLE

What is the pOH of a solution whose pH is 3.0?

Substituting 3.0 for pH in the expression

$$pH + pOH = 14.0$$

gives

$$3.0 + pOH = 14.0$$
$$pOH = 14.0 - 3.0$$
$$pOH = 11.0$$

Indicators

Some **indicators** can be used to determine pH because of their color changes somewhere along this pH scale. Some common indicators and their respective color changes are given below in Table 11.4.

TABLE 11.4 pH Indicators

Indicator	pH Range of Color Change	Color below Lower pH	Color above Higher pH
Methyl orange	3.1–4.4	Red	Yellow
Bromthymol blue	6.0–7.6	Yellow	Blue
Litmus	4.5–8.3	Red	Blue
Phenolphthalein	8.3–10.0	Colorless	Red/Pink

> **TIP**
> Notice that each indicator has its own range of color change.

Here is an example of how to read this chart: At pH values below 4.5, litmus is red; above 8.3, it is blue. Between these values, it is a mixture of the two colors.

Titration-Volumetric Analysis

Knowledge of the concentrations of solutions and the reactions they take part in can be used to determine the concentrations of "unknown" solutions or solids. The use of volume measurement in solving these problems is called **titration**.

A common example of a titration uses acid-base reactions. If you are given a base of known concentration, that is, a standard solution, let us say 0.10 M NaOH, and you want to determine the concentration of an HCl solution, you could **titrate** the solutions in the following manner.

First, introduce a measured quantity, 25.0 milliliters, of the NaOH into a flask by using a pipet or burette in a setup like the one in Figure 11.5. Next, introduce 2 drops of a suitable indicator. Because NaOH and HCl are considered a strong base and a strong acid, respectively, an indicator that changes color in the middle pH range would be appropriate. Litmus solution would be one choice. It is blue in a basic solution but changes to red when the solution becomes acidic. Slowly introduce the HCl until the color change occurs. This point is called the **end point**. The point at which enough acid is added to neutralize all the standard solution in the flask is called the **equivalence point**.

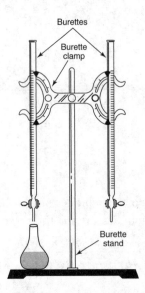

FIGURE 11.5 Burette Setup for Titration

Suppose 21.5 milliliters of HCl was needed to produce the color change. The reaction that occurred was

$$H^+(aq) + OH^-(aq) \rightarrow H_2O(\ell)$$

until all the OH^- was neutralized; then the excess H^+ caused the litmus paper to change color.

To solve the question of the concentration of NaOH, this equation is used:

$$M_{acid} \times V_{acid} = M_{base} \times V_{base}$$

Substituting the known amounts in this equation gives:

$$x\, M_{acid} \times 21.5\ \text{mL} = 0.1\ \text{M} \times 25.0\ \text{mL}$$
$$x = 0.116\ \text{M}$$

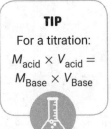

TIP

For a titration:

$M_{acid} \times V_{acid} = M_{Base} \times V_{Base}$

In choosing an indicator for a titration, we need to consider whether the solution formed when the end point is reached has a pH of 7. Depending on the types of acid and base used, the resulting hydrolysis of the salt formed may cause the solution to be slightly acidic, slightly basic, or neutral. If a strong acid and a strong base are titrated, the end point will be at pH 7, and practically any indicator can be used because adding 1 drop of either reagent will change the pH at the end point by about 6 units. For titrations of strong acids and weak bases, we need an indicator, such as methyl orange, that changes color between 3.1 and 4.4 in the acid region. When titrating a weak acid and a strong base, we should use an indicator that changes in the basic range. Phenolphthalein is the suitable choice for this type of titration because it changes color in the pH 8.3 to 10.0 range.

The process of the neutralization reaction can be represented by a titration curve like the one in Figure 11.6 below, which shows the titration of a strong acid with a strong base.

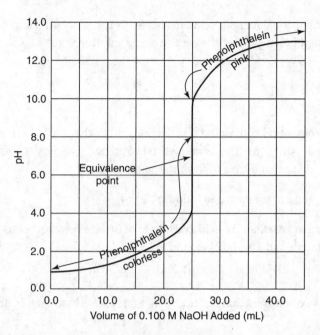

FIGURE 11.6 Titration of a Strong Acid with a Strong Base

EXAMPLE

Find the concentration of acetic acid in vinegar if 21.6 milliliters of 0.20 M NaOH is needed to titrate a 25-milliliter sample of the vinegar.

Solution

Using the equation $M_{acid} \times V_{acid} = M_{base} \times V_{base}$, we have:

$$x\, M_{acid} \times 25\text{ mL} = 0.20 \times 21.6\text{ mL}$$

$$x = \underline{\underline{0.17\, M_{acid}}}$$

Buffer Solutions

Buffer solutions are equilibrium systems that resist changes in acidity and maintain constant pH when acids or bases are added to them. A typical laboratory buffer can be prepared by mixing equal molar quantities of a weak acid such as $HC_2H_3O_2$ and its salt, $NaC_2H_3O_2$. When a small amount of a strong base such as NaOH is added to the buffer, the acetic acid reacts (and consumes) most of the excess OH^- ion. The OH^- ion reacts with the H^+ ion from the acetic acid, thus reducing the H^+ ion concentration in this equilibrium:

$$HC_2H_3O_2(aq) \rightleftharpoons H^+(aq) + C_2H_3O_2^{-}(aq)$$

This reduction of H^+ causes a shift to the right, forming additional $C_2H_3O_2^-$ ions and H^+ ions. For practical purposes, each mole of OH^- added consumes 1 mole of $HC_2H_3O_2$ and produces 1 mole of $C_2H_3O_2^-$ ions.

When a strong acid such as HCl is added to the buffer, the H^+ ions react with the $C_2H_3O_2^-$ ions of the salt and form more undissociated $HC_2H_3O_2$. This does not alter the H^+ ion concentration. Proportional increases and decreases in the concentrations of $C_2H_3O_2^-$ and $HC_2H_3O_2$ do not significantly affect the acidity of the solution.

Salts

A **salt** is an ionic compound containing positive ions other than hydrogen ions and negative ions other than hydroxide ions. The usual method of preparing a particular salt is by neutralizing the appropriate acid and base to form the salt and water.

Five methods for preparing salts are as follows:

TIP

There are five methods for preparing salts.

1. *Neutralization reaction.* An acid and a base neutralize each other to form the appropriate salt and water. For example:

$$2HCl(aq) + Ca(OH)_2(aq) \rightarrow CaCl_2(aq) + 2H_2O(\ell)$$
$$\text{acid} \quad + \quad \text{base} \quad \rightarrow \quad \text{salt} \quad + \quad \text{water}$$

2. *Single replacement reaction.* An active metal replaces hydrogen in an acid. For example:

$$Mg(s) + H_2SO_4(aq) \rightarrow MgSO_4(aq) + H_2(g)$$

3. *Direct combination of elements.* An example of this method is the combination of iron and sulfur. In this reaction, small pieces of iron are heated with powdered sulfur:

$$Fe(s) + S(s) \rightarrow FeS(s)$$
$$\text{iron(II)sulfide}$$

4. *Double replacement.* When solutions of two soluble salts are mixed, they form an insoluble salt compound. For example:

$$AgNO_3(aq) + NaCl(aq) \rightarrow NaNO_3(aq) + AgCl(s)$$

5. *Reaction of a metallic oxide with a nonmetallic oxide.* For example:

$$MgO(s) + SiO_2(s) \rightarrow MgSiO_3(s)$$

Amphoteric Substances

Some substances, such as the HCO_3^- ion, the HSO_4^- ion, the H_2O molecule, and the NH_3 molecule, can act as either proton donors (acids) or proton receivers (bases), depending upon which other substances they come into contact with. These substances are said to be **amphoteric**. Amphoteric substances donate protons in the presence of strong bases and accept protons in the presence of strong acids.

Examples are the reactions of the bisulfate ion, HSO_4^-:

With a strong acid, HSO_4^- accepts a proton:

$$HSO_4^-(aq) + H^+(aq) \rightarrow H_2SO_4(aq)$$

With a strong base, HSO_4^- donates a proton:

$$HSO_4^-(aq) + OH^-(aq) \rightarrow H_2O(\ell) + SO_4^{2-}(aq)$$

CHAPTER 12

Oxidation-Reduction

Learning Objectives

In this chapter, you will learn how to:

o Assign oxidation states to elements in compounds.

o Describe the process of oxidation and reduction.

o Recognize when a substance is being oxidized or reduced.

o Apply the appropriate terms to substances involved in redox reactions.

o Use the concepts of oxidation-reduction to better describe combustion reactions.

A single replacement reaction is characterized by a compound reacting with an element producing a new compound and a new element. Reactions such as these can be predicted to occur (i.e., be spontaneous) if the heat of formation of the compound in the products is negative and reasonably bigger than that of the compound in the reactants. These situations produce exothermic reactions, those in which the system under study becomes lower in energy, which nature tends to want to see occur.

Delving deeper into single replacement reactions allows one to understand why the energy of the system goes down in terms of the chemical process that is occurring. Take the reaction between a solution of silver nitrate and copper metal.

$$2AgNO_3(aq) + Cu(s) \rightarrow Cu(NO_3)_2(aq) + 2Ag(s)$$

The solution of silver nitrate is actually composed of silver ions (Ag^+) and nitrate ions (NO_3^-) surrounded by water molecules in solution. Likewise, the solution of copper(II) nitrate in the products contains ions of Cu^{2+} and NO_3^-. The nitrate ions, then, are found in both the reactants and the products. They can be referred to as **spectator ions**. Because spectator ions can be canceled out, the reaction can be simplified and written as the net ionic equation:

$$2Ag^+(aq) + Cu(s) \rightarrow Cu^{2+}(aq) + 2Ag(s)$$

Analysis of the reaction shown in Figure 12.1 shows that for the process to occur, the positively charged silver ion in the reactants must turn into a silver atom in the products. Likewise, the copper atom must become a positively charged copper ion. How do these changes occur? The answer is simple and represents a major driving force for chemical change—the **transfer of electrons** from a substance that wants them less to a substance that wants them more.

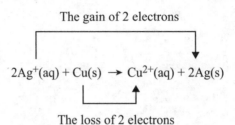

The gain of 2 electrons

$$2Ag^+(aq) + Cu(s) \rightarrow Cu^{2+}(aq) + 2Ag(s)$$

The loss of 2 electrons

FIGURE 12.1 Transfer of Elections

REMEMBER

"Leo the lion says Ger" stands for Loss of Electrons is Oxidation and Gain of Electrons is Reduction.

Whenever the transfer of electrons occurs during a chemical process, one substance must lose electrons in order for the other to gain them. The loss of electrons has a name in chemistry: **oxidation**. Accordingly, the gaining of electrons is named too: **reduction**. Since one process can't happen without the other, these processes are often intertwined into one distinctly chemical term called **redox**. Redox reactions are those in which oxidation and reduction occur in complementary ways. Redox is a major reaction category in chemistry. All single replacement reactions are redox reactions; however, all redox reactions are not single replacement reactions. Before we move on to other reaction types that can be seen as redox, let's take a further look at the single replacement category from a redox perspective.

The activity series of metals, the use of which is another way to predict the spontaneity of single replacement reactions, can now be seen as a measure of the desire of certain metals to lose electrons compared to other metals. In the single replacement reaction example previously discussed, the process can be said to occur because the copper atom has a greater desire to lose electrons than does the silver atom. In other words, the reaction will proceed in the forward direction as written (as opposed to going backward) because there is a natural push for copper to lose electrons more than for the silver to lose electrons. Elements higher on the activity series then are simply those that have a stronger desire to lose electrons than the ones below them.

Oxidation States

Because the transfer of electrons in chemical reactions is so prevalent, chemists have developed a manner to keep track of their movement. It's important to note that for processes involving ionic species, of the type we have looked at so far, it is somewhat straightforward to distinguish redox from non-redox: The sign and magnitude of the charge on the ions involved allows for an easy view of what is taking place. Using the spontaneous reaction between zinc and iron(III) chloride referenced in the previous section, the zinc started as an atom with no net charge and turned into an ion with a 2+ charge. During chemical reactions, in order for the charge on particles to increase, negative electrons have to be lost and the process of oxidation occurs. A similar analysis of the iron in the reaction shows that the iron began as a 3+ ion (ferric) and ended as a neutral iron atom in the products. In other words, the iron went *down* in charge. Particles that go down in charge do so by gaining negative electrons. This description gives insight to why the process is referred to as *reduction* despite the fact that electrons are being *gained*. Reduction refers to the *change in charge* involving ionic species, not to the change in the number of electrons possessed by a particle as a result of the process taking place. Since reduction involves a decrease in charge, oxidation must involve an increase in charge (when ionic species are involved). Using the term *oxidation* may seem like a strange way to describe this process, but its use will be better understood from a historical context when **combustion** reactions are looked at as redox processes in an upcoming section of this chapter.

As discussed, reactions involving ions are relatively simple to recognize as redox (or not) because it is straightforward to identify changes in charges if they occur. It's not quite so simple when redox reactions involve molecular species. For these types of reactions, as well as those involving ionic species, a related but noteworthily different term is used to describe the responsibility particles have for the electrons around them as a chemical reaction unfolds. To keep track of the transfer of electrons in all formulas (ionic or molecular), chemists have devised a system of electron bookkeeping called **oxidation states** (or **oxidation numbers**). In this system, an oxidation state is assigned to each member of a formula using rules that recognize the degree to which electrons *practically belong* to a particular element in a given substance from an ionic bonding perspective. Basically, elements with high electronegativities are given responsibility for the electrons in a bond, ionic or covalent, and the change in this responsibility will be noted by changes in their oxidation states. In this regard, the oxidation states system assumes an ionic perspective for *all* bonding, i.e., electrons are not shared but *belong* to one element or the other in a chemical bond. Although we know that electrons *are* shared in a large number of chemical bonds, particularly those described as covalent (or molecular), assigning responsibility for the electrons in this way allows for easy recognition of how the accountability for electrons changes during all chemical processes.

Oxidation states are designated by a small number superscript *preceded* by a plus or minus sign. This is not to be confused with the ionic charges we have been using thus far that are shown as plus or minus signs *to the right* of the magnitude of ionic charge as a superscript.

The Rules for Assigning an Oxidation State

The basic rules for assigning an oxidation state to an element in a substance's formula are given below. By applying these simple rules, you can assign oxidation states to elements in practically all substances you may encounter as a beginning chemistry student. To apply these rules, remember that the **sum of the oxidation states must be zero for an electrically neutral compound**. For a polyatomic ion, the **sum of the oxidation states must be equal to the charge on the ion**.

1. The oxidation state of an atom in an element is zero. Examples: 0 for $Na(s)$, $O_2(g)$, and $Hg(\ell)$.

2. The oxidation state of a monatomic ion is the same as its charge. Examples: $+1$ for Na^+ and -1 for Cl^-.

3. The oxidation state for fluorine is -1 in its compounds. Examples: HF, hydrogen fluoride, has one H at $+1$ and one F at -1. PF_3 has one P at $+3$ and three F's at -1 each. (Note that in each compound the sum of the oxidation states is equal to zero.)

4. The oxidation state of oxygen is usually -2 in its compounds. Example: H_2O has two H's at $+1$ each and one O at -2. (Exceptions occur when the oxygen is bonded to fluorine and the oxidation state of fluorine takes precedence, and in peroxide compounds where the oxidation state is assigned the value of -1.)

5. The oxidation state of hydrogen in most compounds is +1. Examples: H_2O, HCl, and NH_3. (In hydrides, where hydrogen acts like an anion compounded with a metal, there is an exception, however. In this case, hydrogen is assigned the value of -1. Examples: LiH and KH.)

An example of determining the oxidation states of other elements in chemical formulas follows.

EXAMPLE

In Na_2SO_4, what is the oxidation state of sulfur?

The first thing to recognize is that this compound is an ionic substance. Ionic substances have two parts—the cation and the anion. In this case, the cation is monatomic and the anion is polyatomic. The cation is Na^+ and the anion is SO_4^{2-} (sulfate). By Rule #2, the oxidation state of the sodium is +1 because the oxidation states of monatomic ions are the same as their charges. By Rule #4, the oxidation state of the oxygen in the sulfate is -2. Now you can look at the complete formula and calculate the oxidation state of the sulfur.

$$Na_2SO_4$$

There are two sodium ions.   There are four oxygen atoms.
$2 \times (+1) = +2$ $4 \times (-2) = -8$

Since the positive sum and the negative sum must equal 0,

$$(+2) + x + (-8) = 0$$

the sulfur must have a +6 oxidation state.

Using Oxidation States to Recognize Redox Reactions

Once oxidation states can be assigned to elements in the substances involved in a chemical reaction, recognition of the process as being redox or not is straightforward. If the oxidation states change, then a transfer of electrons is taking place. If the oxidation states remain the same, then a redox reaction is not occurring.

Consider these two reactions:

$$Na_2S(aq) + CuSO_4(aq) \rightarrow Na_2SO_4(aq) + CuS(s)$$
and
$$2Na(\ell) + Cl_2(g) \rightarrow 2NaCl(s)$$

The oxidation states of each element in all the substances can be determined as shown above each of them here:

$$\overset{+1\ -2}{Na_2S}(aq) + \overset{+2\ +6\ -2}{CuSO_4}(aq) \rightarrow \overset{+1\ +6\ -2}{Na_2SO_4}(aq) + \overset{+2\ -2}{CuS}(s)$$

and

$$\overset{0}{2Na}(\ell) + \overset{0}{Cl_2}(g) \rightarrow \overset{+1\ -1}{2NaCl}(s)$$

Because the oxidation states of the elements in the first reaction don't change, this reaction is not a redox process. This is a precipitation reaction. The second reaction does exhibit a change in oxidation states and should be viewed as a redox reaction. In this reaction, the sodium is being oxidized and the chlorine is being reduced. The chlorine could not be reduced (i.e., gain electrons) if the sodium wasn't being oxidized (i.e., losing electrons). In one regard, the sodium is acting as a facilitator for the reduction of the chlorine. As such, it is referred to as a **reducing agent**. Likewise, the sodium would not lose its electron if it had nowhere to go, so the chlorine is referred to as the **oxidizing agent** in this reaction. Oxidizing agents, then, contain elements that are capable of being reduced by other substances (reducing agents) that contain elements that are capable of being oxidized. Sometimes the terms **oxidizer** and **reducer** are used as labels on the bottles of substances with the tendency to act as oxidizing and reducing agents, respectively.

EXAMPLE

Identify the elements that are being oxidized and reduced in the following reaction. Also, name the oxidizing and reducing agents.

$$2H_2O(\ell) + 2MnO_4^-(aq) + I^-(aq) \rightarrow 2MnO_2(aq) + IO_3^-(aq) + 2OH^-(aq)$$

The oxidation states of the hydrogen and oxygen are not changing in the reaction. The oxidation states of the manganese and the iodine are changing, as shown below.

$$\overset{+7}{\downarrow} \qquad \overset{-1}{\downarrow} \qquad \overset{+4}{\downarrow} \qquad \overset{+5}{\downarrow}$$

$$2H_2O(\ell) + 2MnO_4^-(aq) + I^-(aq) \rightarrow 2MnO_2(aq) + IO_3^-(aq) + 2OH^-(aq)$$

Since the manganese is changing from an oxidation state of $+7$ to that of $+4$, the manganese is being reduced. The iodine is being oxidized from a value of -1 to a value of $+5$. Manganese is gaining electron responsibility while iodine is losing it. In that light, then, the permanganate ion (MnO_4^-), which contains the manganese, is facilitating the process in which iodine is being oxidized and so is referred to as the oxidizing agent. A typical source of the permanganate ion in chemical reactions like this one is from the compound potassium permanganate, $KMnO_4$, whose bottle is generally labeled with the term "oxidizer." Along a line of similar thinking, the iodide ion, I^-, can be called the reducing agent in this reaction.

Combustion Reactions

Combustion reactions are those chemical processes in which substances (called *fuels*) are rapidly oxidized, accompanied by the release of heat and usually light. Combustion is also referred to as **burning**. Typical, common, and historically well-known combustion reactions involve using oxygen as the oxidizing agent—hence, the name for the process of losing electrons has became known as *oxidation*. When a combustion reaction is complete, the elements in the burning fuel form compounds with the oxidizing agent and the responsibility for electrons changes.

The reaction between magnesium and oxygen is a common example of combustion. When the reaction occurs

$$2Mg(s) + O_2(g) \rightarrow 2MgO(s)$$

the oxidation states of magnesium and oxygen change from 0 each to +2 and −2, respectively. A blinding light and large quantity of heat is released by the system as the reaction unfolds. The amount of heat released when 1 mole of a fuel burns is referred to as its *heat of combustion* and is symbolized ΔH_c. The ΔH_c for Mg is 602 kJ/mol. The *change in enthalpy* associated with the burning of 1 mole of carbon

$$C(s) + O_2(g) \rightarrow CO_2(g) + 393.5 \text{ kJ}$$

is −393.5 kJ, also written as $\Delta H_c = -393.5$ kJ, and it represents the heat released in another combustion/redox reaction. The heats of combustion for many common substances can easily be found in reference tables.

Hydrocarbon Fuel Combustion Reactions

Another common combustion reaction involves the burning of hydrocarbon fuels. Methane, CH_4, is a typical hydrocarbon fuel. It is the main component of **natural gas**. Analysis of the reaction

oxidation states: −4   0   +4 −2

$$CH_4(g) + 2O_2(g) \rightarrow CO_2(g) + 2H_2O(\ell)$$

shows that carbon is being oxidized and the oxygen is being reduced. This fuel is used to heat homes and cook food and should be familiar to all chemistry students as the gas used in their laboratory burners.

EXAMPLE

Describe the reaction that occurs when propane, C_3H_8 (the hydrocarbon fuel used in backyard barbecues) combusts, releasing 2,221 kJ when 1 mole is burned.

It should be noted that when hydrocarbon fuels combust completely in the presence of oxygen, the products of the reaction are carbon dioxide and water. Therefore, the reaction is:

oxidation states: −8/3*   0   +4 −2

$$C_3H_8(g) + 5O_2(g) \rightarrow 3CO_2(g) + 4H_2O(\ell) + 2,221 \text{ kJ}$$

Take note that oxidation states may be expressed in fractions.

Once again, the carbon is oxidized from the fraction −8/3 to +4 (increasing from negative to positive) and the oxygen is reduced from 0 to −2. Oxygen is the oxidizing agent and propane is the reducing agent. When this reaction occurs, light and heat are released and the heat of combustion, ΔH_c, is −2,221 kJ.

CHAPTER 13

Carbon and Organic Chemistry

Learning Objectives

In this chapter, you will learn how to:

○ Describe the bonding patterns of carbon and its allotropic forms.

○ Explain the structural pattern and naming of the alkanes, alkenes, and alkynes, and their isomers.

○ Show graphically how hydrocarbons can be changed and the development of these functional groups, their structures, and their names: alcohols, aldehydes, ketones, esters, and amines.

Carbon is unique. It forms inorganic substances such as carbon dioxide, graphite, and diamonds. It also forms organic substances without which life could not exist. It forms planar substances, tetrahedrons, and rings.

Forms of Carbon

The element carbon occurs mainly in three allotropic forms: diamond, graphite, and amorphous (although some evidence shows the amorphous forms have some crystalline structure). In the mid-1980s, **fullerenes** were identified as a new allotropic form of carbon. They are found in soot that forms when carbon-containing materials are burned with limited oxygen. Their structure consists of near-spherical cages of carbon atoms resembling geodesic domes.

The **diamond** form has a close-packed crystal structure that gives it its property of extreme hardness. In it each carbon is bonded to four other carbons in a tetrahedron arrangement as seen in Figure 13.1. These covalent solids form crystals that can be viewed as a single giant molecule made up of an almost endless number of covalent bonds. Because all of the bonds in this structure are equally strong, covalent solids are often very hard, and they are notoriously difficult to melt. Diamond is the hardest natural substance. At atmospheric pressure, it melts at 3,550°C.

> **TIP**
> Allotropic forms
> of carbon:
> diamond
> graphite
> amorphous
> fullerenes

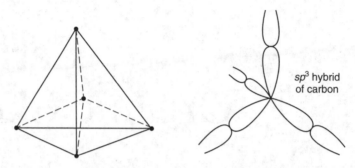

FIGURE 13.1 Diamond Structure

It has been possible to make synthetic diamonds in machines that subject carbon to extremely high pressures and temperatures. Most of these diamonds are used for industrial purposes, such as dies.

The graphite form is made up of planes of hexagonal structures that are weakly bonded to the planes above and below. This explains graphite's slippery feeling and makes it useful as a dry lubricant. Graphite is also mixed with clay to make "lead" for lead pencils. Its structure can be seen below in Figure 13.2. Graphite also has the property of being an electrical conductor.

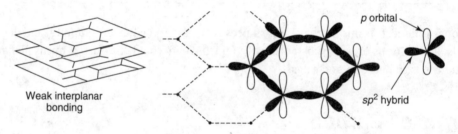

FIGURE 13.2 Graphite Structure

Some common amorphous forms of carbon are charcoal, coke, bone black, and lampblack.

Carbon Dioxide

Carbon dioxide (CO_2) is a widely distributed gas that makes up 0.04% of the air. There is a cycle that keeps this figure relatively stable. It is shown in Figure 13.3.

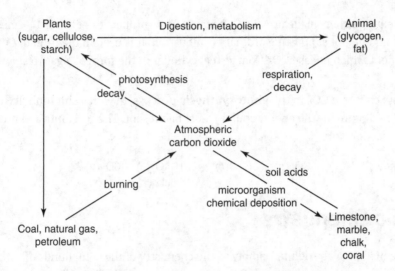

FIGURE 13.3 Carbon Dioxide Cycle

Laboratory Preparation of CO_2

The usual laboratory preparation consists of reacting calcium carbonate (marble chips) with hydrochloric acid, although any carbonate or bicarbonate and any common acid could be used. The gas is collected by water displacement or air displacement.

The test for carbon dioxide consists of passing it through limewater ($Ca(OH)_2$). If CO_2 is present, the limewater turns cloudy because of the formation of a white precipitate of finely divided $CaCO_3$:

$$Ca(OH)_2(aq) + CO_2(g) \rightarrow CaCO_3(s) + H_2O(\ell)$$

Continued passing of CO_2 into the solution will eliminate the cloudy condition because the insoluble $CaCO_3$ becomes soluble calcium bicarbonate ($Ca(HCO_3)_2$):

$$CaCO_3(s) + H_2O(\ell) + CO_2(g) \rightarrow Ca^{2+}(HCO_3^-)^2(aq)$$

This reaction can easily be reversed with increased temperature or decreased pressure. This is the way stalagmites and stalactites form on the floors and roofs of caves, respectively. The ground water containing calcium bicarbonate is deposited on the roof and floor of the cave and decomposes into solid calcium carbonate formations.

Important Uses of CO_2

1. Because CO_2 is the acid anhydride of carbonic acid, it forms the acid when reacted with soft drinks, thus making them "carbonated" beverages.

$$CO_2(g) + H_2O(\ell) \rightarrow H_2CO_3(aq)$$

2. Solid carbon dioxide ($-78°C$), or "dry ice," is used as a refrigerant because it has the advantage of not melting into a liquid; instead, it sublimes and in the process absorbs 3 times as much heat per gram as ice.

3. Fire extinguishers make use of CO_2 because of its properties of being $1\frac{1}{2}$ times heavier than air and not supporting ordinary combustion. It is used in the form of CO_2 extinguishers, which release CO_2 from a steel cylinder in the form of a gas to smother the fire.

4. Plants consume CO_2 in the **photosynthesis process**, in which chlorophyll (the catalyst) and sunlight (the energy source) must be present. The reactants and products of this reaction are:

$$6CO_2(g) + 6H_2O(\ell) \rightarrow C_6H_{12}O_6(s) + 6O_2(g)$$
cellulose

Organic Chemistry

Organic chemistry may be defined simply as the chemistry of the compounds of carbon. Since Friedrich Wöhler synthesized urea in 1828, chemists have synthesized thousands of carbon compounds in areas of dyes, plastic, textile fibers, medicines, and drugs. The number of organic compounds has been estimated to be in the neighborhood of a million and constantly increasing.

The carbon atom (atomic number 6) has four electrons in its outermost energy level, which show a tendency to be shared (electronegativity of 2.5) in covalent bonds. By this means, carbon bonds to other carbons, hydrogens, halogens, oxygen, and other elements to form the many compounds of organic chemistry.

Hydrocarbons

Hydrocarbons, as the name implies, are compounds containing only carbon and hydrogen in their structures. The simplest hydrocarbon is methane, CH_4. As previously mentioned, this type of formula, which shows the kinds of atoms and their respective numbers, is called an **empirical** formula. In organic chemistry, this is not sufficient to identify the compound it is used to represent. For example, the empirical formula C_2H_6O could denote either an ether or an ethyl alcohol. For this reason, a **structural** formula is used to indicate how the atoms are arranged in the molecule. The ether of C_2H_6O looks like this:

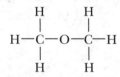

whereas the ethyl alcohol is represented by this structural formula:

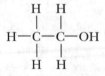

TIP
Organic chemistry makes use of structural formulas to show atomic arrangements.

To avoid ambiguity, structural formulas are more often used than empirical formulas in organic chemistry. The structural formula of methane is:

$$\begin{array}{c} \text{H} \\ | \\ \text{H}\!-\!\text{C}\!-\!\text{H} \\ | \\ \text{H} \end{array}$$

Alkane Series (Saturated)

Methane is the first member of a hydrocarbon series called the **alkanes** (or paraffin series). The general formula for this series is C_nH_{2n+2}, where n is the number of carbons in the molecule. Table 13.1 provides some essential information about this series. Since many other organic structures use the stem of the alkane names, you should learn these names and structures well. Notice that, as the number of carbons in the chain increases, the boiling point also increases. The first four alkanes are gases at room temperature; the subsequent compounds are liquid, then become more viscous with increasing length of the chain.

Since the chain is increased by a carbon and two hydrogens in each subsequent molecule, the alkanes are referred to as a **homologous** series.

The alkanes are found in petroleum and natural gas. They are usually extracted by fractional distillation, which separates the compounds by varying the temperature so that each vaporizes at its respective boiling point.

When the alkanes are burned with sufficient air, the compounds formed are CO_2 and H_2O. An example is:

$$2C_2H_6(g) + 7O_2(g) \rightarrow 4CO_2(g) + 6H_2O(g)$$

The alkanes can be reacted with halogens so that hydrogens are replaced by a halogen atom: These are called **alkyl halides**.

$$\begin{array}{ccc} \text{H} & & \text{H} \\ | & & | \\ \text{H}\!-\!\text{C}\!-\!\text{H} + \text{Br}_2 \rightarrow & \text{H}\!-\!\text{C}\!-\!\text{Br} & + \text{HBr} \\ | & & | \\ \text{H} & & \text{H} \end{array}$$

Some common substitution compounds of methane are:

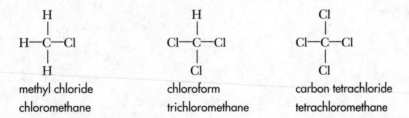

methyl chloride   chloroform   carbon tetrachloride
chloromethane   trichloromethane   tetrachloromethane

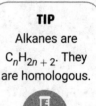

TIP

Alkanes are C_nH_{2n+2}. They are homologous.

TIP

Learn the names of the first ten alkanes.

Table 13.1 The Alkanes

IUPAC Name	Molecular Formula	Number of Structural Isomers	Structure	State at Room Temperature	Boiling Point (°C)
Methane	CH_4	1		Gas	−162
Ethane	C_2H_6	1		Gas	−89
Propane	C_3H_8	1		Gas	−42
n-Butane	C_4H_{10}	2		Gas	0
n-Pentane	C_5H_{12}	3		Liquid (Note: Solid at 17 carbons in the chain)	36
n-Hexane	C_6H_{14}	5	$CH_3-CH_2-CH_2-CH_2-CH_3$	Liquid	69
n-Heptane	C_7H_{16}	7	$CH_3-CH_2-CH_2-CH_2-CH_2-CH_3$	Liquid	98
n-Octane	C_8H_{18}	18	$CH_3-CH_2-CH_2-CH_2-CH_2-CH_2-CH_3$	Liquid	126
n-Nonane	C_9H_{20}	35	$CH_3-CH_2-CH_2-CH_2-CH_2-CH_2-CH_2-CH_3$	Liquid	151
n-Decane	$C_{10}H_{22}$	75	$CH_3-CH_2-CH_2-CH_2-CH_2-CH_2-CH_2-CH_2-CH_3$	Liquid	174

Naming Alkane Substitutions

When an alkane hydrocarbon has an end hydrogen removed, it is referred to as an alkyl substituent or group. The respective name of each is the alkane name with *-ane* replaced by *-yl*. These are called **alkyl groups**. Use of the methyl and butyl group are shown below in Figure 13.4.

TIP

Replace *-ane* with *-yl* to form alkyl groups.

Alkane	Alkyl Group	Compounds
methane	methyl	bromomethane
H‑C‑H	H‑C‑	H‑C‑Br
butane	butyl	1-chlorobutane
H‑C‑C‑C‑C‑H	H‑C‑C‑C‑C‑	H‑C‑C‑C‑C‑Cl

FIGURE 13.4 Alkyl Groups

One method of naming a substitution product is to use the alkyl name for the respective chain and the halide as shown above. The halogen takes the form of *fluoro-*, *bromo-*, *iodo-*, and so on, depending on the halogen, and is attached to an alkane name. It precedes the alkane name, as shown above in bromomethane and 1-chlorobutane.

The IUPAC system uses the name of the longest carbon chain as the parent chain. The carbon atoms are numbered in the parent chain to indicate where branching or substitution takes place. The direction of numbering is chosen so that the lowest numbers possible are given to the side chains. The complete name of the compound is arrived at by first naming the attached group, each of these being prefixed by the number of the carbon to which it is attached, and then the parent alkane. If a particular group appears more than once, the appropriate prefix (*di*, *tri*, and so on) is used to indicate how many times the group appears. A carbon atom number must be used to indicate the position of each such group. If two or more of the same group are attached to the same carbon atom, the number of the carbon atom is repeated. If two or more different substituted groups are in a name, they are arranged alphabetically.

EXAMPLE

2,2-dimethylbutane

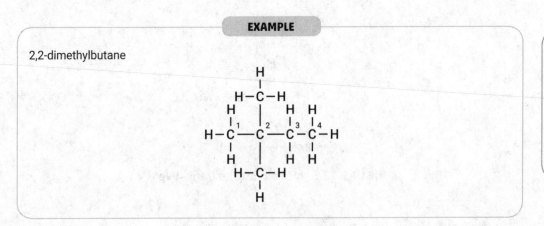

TIP

Numbers have been added to the longest chain for identification only.

EXAMPLE

1,1-dichloro-3-ethyl-2,4-dimethylpentane

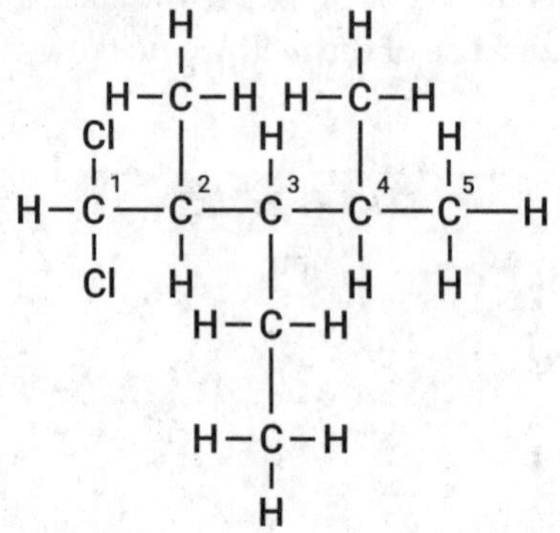

Cycloalkanes

Starting with propane in the alkane series, it is possible to get a ring form by attaching the two chain ends. This reduces the number of hydrogens by two.

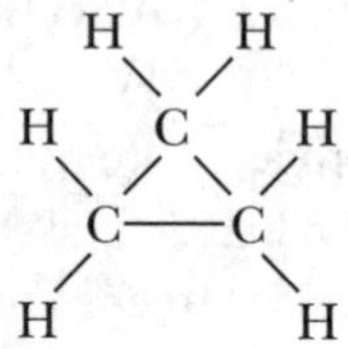

This hydrocarbon is called **cyclopropane**.

Cycloalkanes are named by adding the prefix *cyclo-* to the name of the straight-chain alkane with the same number of hydrocarbons, as shown above.

When there is only one alkyl group attached to the ring, no position number is necessary. When there is more than one alkyl group attached to the ring, the carbon atoms in the ring are numbered to give the lowest numbers possible to the alkyl groups. This means that one of the alkyl groups will always be in position 1. The general formula is C_nH_{2n}.

An example of a cycloalkane derivative is shown in Figure 13.5.

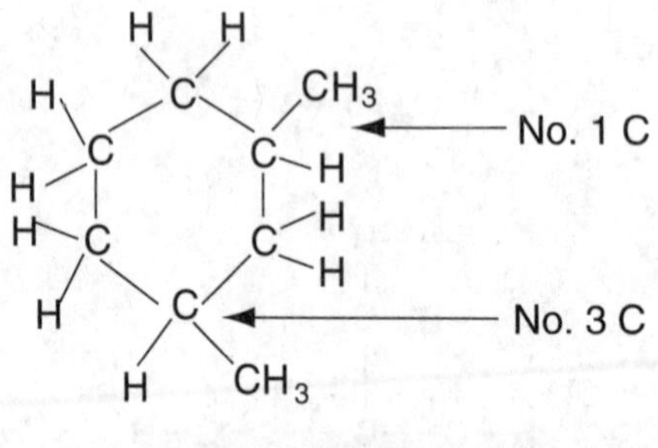

No. 1 C

No. 3 C

1,3-dimethylcyclohexane

FIGURE 13.5 A Cyclohexane Derivative

If there are two or more alkyl groups attached to the ring, number the carbon atoms in the ring. Assign position number one to the alkyl group that comes first in alphabetical order, then number in the direction that gives the rest of the alkyl groups the lowest numbers possible.

Because all the members of the alkane series have single covalent bonds, this series and all such structures are said to be **saturated**.

If the hydrocarbon molecule contains double or triple covalent bonds, it is referred to as **unsaturated**.

Properties and Uses of Alkanes

Properties for some straight-chain alkanes are indicated in Table 13.1. The trends in these properties can be explained by examining the structures of alkanes. The carbon-hydrogen bonds are nonpolar. The only forces of attraction between nonpolar molecules are weak intermolecular, or London dispersion, forces. These forces increase as the mass of a molecule increases.

The table also shows the physical states of alkanes. Smaller alkanes exist as gases at room temperature, while larger ones exist as liquids. Gasoline and kerosene consist mostly of liquid alkanes. Seventeen carbons are needed in the chain for the solid form to occur. Paraffin wax contains solid alkanes.

The differences in the boiling points of mixtures of the liquid alkanes found in petroleum make it possible to separate the various components by **fractional distillation**. This is the major industrial process used in refining petroleum into gasoline, kerosene, lubricating oils, and several other minor components.

Alkene Series (Unsaturated)

The **alkene** series has a double covalent bond between two adjacent carbon atoms. The general formula of this series is C_nH_{2n}. In naming these compounds, the suffix of the alkane is replaced by -ene. Two examples:

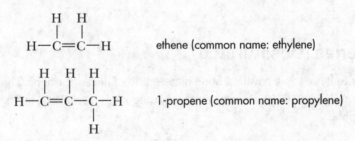

Naming a more complex example is:

$$CH_2=\overset{\overset{\displaystyle CH_2-CH_3}{\displaystyle |}}{C}-CH_2-CH_2-CH_3$$

The position number and name of the alkyl group are in front of the double-bond position number. The alkyl group above is an ethyl group. It is on the second carbon atom of the parent hydrocarbon.

The name is 2-ethyl-1-pentene.

If the double bond occurs on an interior carbon, the chain is numbered so that the position of the double bond is designated by the lowest possible number assigned to the first doubly bonded carbon. For example:

$$
\begin{array}{c}
\quad\ \ \text{H}\ \ \ \text{H}\qquad\quad\ \ \text{H} \\
\quad\ \ |\ \ \ \ | \qquad\quad\ \ | \\
\text{H}-\text{C}-\text{C}-\text{C}=\text{C}-\text{C}-\text{H} \qquad \text{2-pentene} \\
\quad\ \ |\ \ \ \ |\ \ \ \ |\ \ \ \ |\ \ \ \ | \\
\quad\ \ \text{H}\ \ \ \text{H}\ \ \ \text{H}\ \ \ \text{H}\ \ \ \text{H}
\end{array}
$$

The bonding is more complex in the double covalent bond than in the single bonds in the molecule. Using the orbital pictures of the atom seen in Figure 13.6, we can show this as follows:

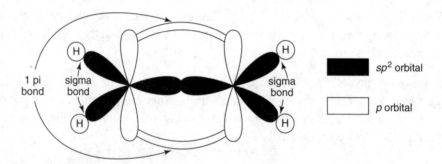

FIGURE 13.6 Sigma and Pi (Double) Bond in Ethane

The two *p* lobes attached above and below constitute *one* bond called a pi (π) bond.

The sp^2 orbital bonds between the carbons and with each hydrogen are referred to as sigma (σ) bonds.

Alkyne Series (Unsaturated)

The **alkyne** series has a triple covalent bond between two adjacent carbons. The general formula of this series is C_nH_{2n-2}. In naming these compounds, the alkane suffix is replaced by -*yne*. Two examples:

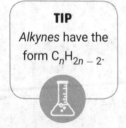

$$\text{H}-\text{C}\equiv\text{C}-\text{H} \qquad \text{ethyne (common name: acetylene)}$$

$$
\begin{array}{c}
\qquad\qquad\ \ \text{H} \\
\qquad\qquad\ \ | \\
\text{H}-\text{C}\equiv\text{C}-\text{C}-\text{H} \qquad \text{propyne} \\
\qquad\qquad\ \ | \\
\qquad\qquad\ \ \text{H}
\end{array}
$$

The orbital structure of ethyne can be shown as follows in Figure 13.7:

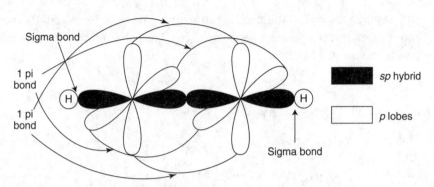

■	*sp* hybrid
□	*p* lobes

TIP
Notice that *pi bonds* are between *p* orbitals and that *sigma bonds* are between *s* and *p* orbitals.

FIGURE 13.7 Orbital Structure of Ethyne

The bonds formed by the *p* orbitals and the one bond between the *sp* orbitals make up the triple bond.

The preceding examples show only one triple bond. If there is more than one triple bond, modify the suffix to indicate the number of triple bonds. For example, 2 would be a diyne, 3 would be a triyne, and so on. Next add the names of the alkyl groups if they are attached. Number the carbon atoms in the chain so that the first carbon atom in the triple bond nearest the end of the chain has the lowest number. If numbering from both ends gives the same positions for two triple bonds, then number from the end nearest the first alkyl group. Then, place the position numbers of the triple bonds immediately before the name of the parent hydrocarbon alkyne and place the alkyl group position numbers immediately before the name of the corresponding alkyl group.

Two more examples of alkynes are:

$$CH_3-CH_2-CH_2-C{\equiv}CH$$

1-pentyne

$$CH{\equiv}C-CH-CH_3$$
$$|$$
$$CH_3$$

2-methyl-1-butyne

Naming a more complex example is:

$$CH_2-CH_3$$
$$|$$
$$CH{\equiv}C-CH_2-CH_2-CH_3$$

The position number and the name of the alkyl group are placed in front of the double-bond position number. The alkyl group above is an ethyl group. It is on the second carbon atom of the parent hydrocarbon.

The name is 2-ethyl-1-pentyne.

Aromatics

The **aromatic** compounds are unsaturated ring structures. The basic formula of this series is C_nH_{2n-6}, and the simplest compound is benzene (C_6H_6). The benzene structure is a resonance structure that is represented like this:

Note: The carbon-to-carbon bonds are neither single nor double bonds but hybrid bonds. This structural representation is called **resonance structures**.

The benzene resonance structure can also be shown in Figure 13.8a:

FIGURE 13.8a Benzene Resonance Structure

The orbital structure can be represented as shown in Figure 13.8b:

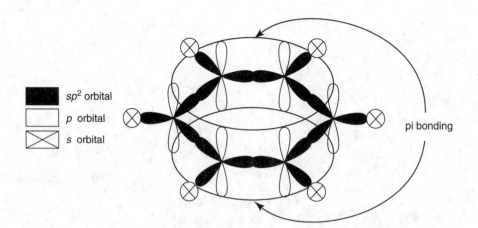

FIGURE 13.8b Benzene Orbital Structure

Most of the aromatics have an aroma, thus the name "aromatic."

The C_6H_5 group is a substituent called phenyl. This is the benzene structure with one hydrogen missing. If the phenyl substituent adds a methyl group, the compound is called toluene or methyl benzene as seen in Figure 13.9.

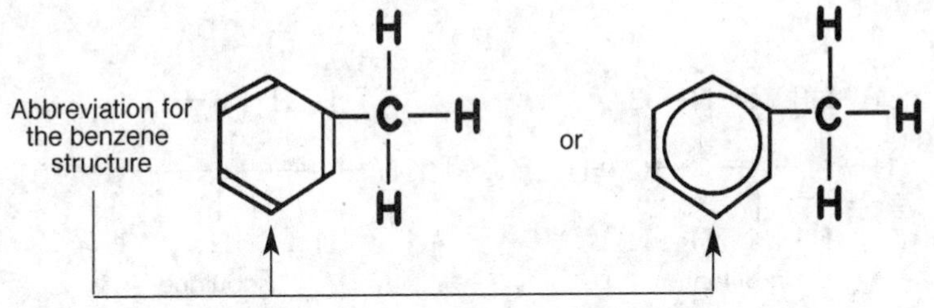

FIGURE 13.9 The Structure of Toluene

Two other members of the benzene series and their structures:

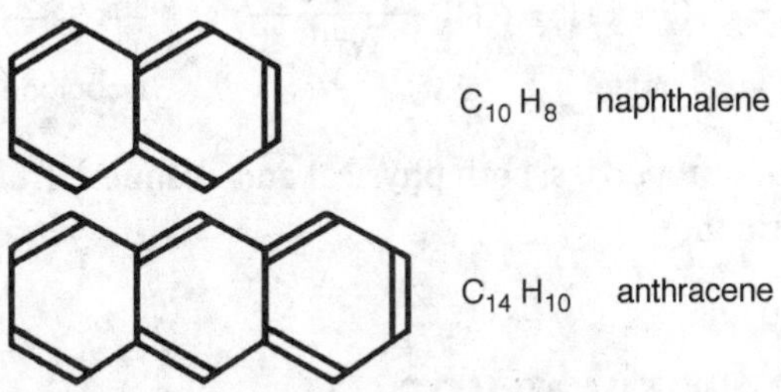

$C_{10}H_8$ naphthalene

$C_{14}H_{10}$ anthracene

FIGURE 13.10 Fused Rings

The IUPAC system of naming benzene derivatives, as with chain compounds, involves numbering the carbon atoms in the ring in order to pinpoint the locations of the side chains. However, if only two groups are substituted in the benzene ring, the compound formed will be a benzene derivative having three possible isomeric forms. In such cases, the prefixes **ortho-**, **meta-**, and **para-**, abbreviated as *o-*, *m-*, and *p-*, may be used to name the isomers. Figure 13.11 shows these positioning possibilities using two methyl groups. In the ortho- structure, the two substituted groups are located on adjacent carbon atoms. In the meta- structure, they are separated by one carbon atom. In the para- structure, they are separated by two carbon atoms.

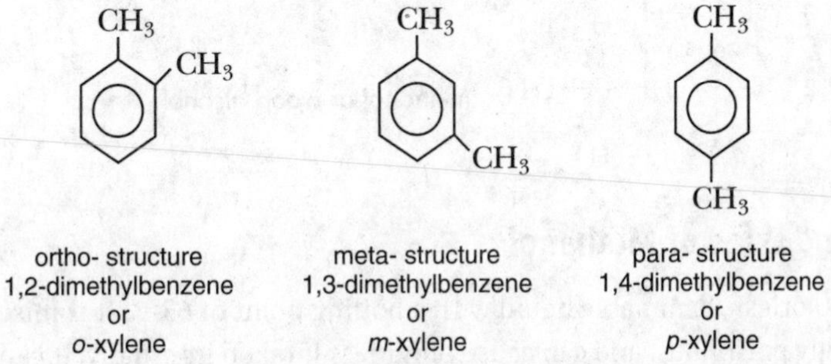

ortho- structure
1,2-dimethylbenzene
or
o-xylene

meta- structure
1,3-dimethylbenzene
or
m-xylene

para- structure
1,4-dimethylbenzene
or
p-xylene

FIGURE 13.11 Ortho-, Meta-, and Para- Diclorobenzene

175

Isomers

Many of the chain hydrocarbons can have the same formula, but their structures may differ. For example, butane is the first compound that can have two different structures or **isomers** for the same formula.

$$
\begin{array}{cccc}
& H & H & H & H \\
& | & | & | & | \\
H- & C- & C- & C- & C-H \\
& | & | & | & | \\
& H & H & H & H
\end{array}
$$

n-butane

isobutane

This isomerization can be shown by the following equation:

$$
CH_3-CH_2-CH_2-CH_3 \xrightarrow[70-100°C]{AlCl_3} CH_3-\overset{\displaystyle CH_3}{\underset{\displaystyle |}{CH}}-CH_3
$$

butane

isobutane

The isomers have different properties, both physical and chemical, from those of hydrocarbons with the normal structure.

Hydrocarbon Derivatives

Hydrocarbon derivatives are compounds containing mostly hydrogen and carbon in which specific groupings of atoms are attached. Many families of hydrocarbon derivatives are familiar, even to non-chemists, and are based on what the specific grouping of attached atoms happen to be.

Alcohols—Methanol and Ethanol

The simplest **alcohols** are alkanes that have one or more hydrogen atoms replaced by the hydroxyl group, –OH. This is called its **functional group**.

Methanol is the simplest alcohol. Its structure is:

$$
\begin{array}{c}
H \\
| \\
H-C-OH \\
| \\
H
\end{array}
$$
methanol or wood alcohol

Properties and Uses of Methanol

Methanol is a colorless, flammable liquid with a boiling point of 65°C. It is miscible with water, is exceedingly poisonous, and can cause blindness if taken internally. It can be used as a fuel, as a solvent, and as a denaturant to make ethyl alcohol, unsuitable for drinking.

Ethanol is the best known and most used alcohol. Its structure is:

$$H-\overset{\overset{\displaystyle H}{|}}{\underset{\underset{\displaystyle H}{|}}{C}}-\overset{\overset{\displaystyle H}{|}}{\underset{\underset{\displaystyle H}{|}}{C}}-OH \quad \text{ethanol}$$

Notice that the alcohol names are derived from the alkanes by replacing the *e* ending with -*ol*.

Its common names are ethyl alcohol and grain alcohol.

Properties and Uses of Ethanol

Ethanol is a colorless, flammable liquid with a boiling point of 78°C. It is miscible with water and is a good solvent for a wide variety of substances (these solutions are often referred to as "tinctures"). It can be used as an antifreeze because of its low freezing point, −115°C, and for making acetaldehyde and ether. It is presently used in gasoline as an alternative to reduce the use of petroleum.

Other Alcohols

Isomeric alcohols have similar formulas but different properties because of their differences in structure. If the −OH is attached to an end carbon, the alcohol is called a primary alcohol. If attached to a "middle" carbon, it is called a secondary alcohol. Some examples:

1-propanol
or
propyl alcohol

← isomers →

2-propanol
or
isopropyl alcohol

The number in front of the name indicates to which carbon the −OH ion is attached.

Other alcohols with more than one −OH group:

ethylene glycol
or
1,2-ethanediol

A colorless liquid, high boiling point, low freezing point. Used as permanent antifreeze in automobiles.

glycerine
or
glycerol
or
1,2,3-propanetriol

Colorless liquid, odorless, viscous, sweet taste. Used to make nitroglycerine, resins for paint, and cellophane.

Aldehydes

The functional group of an **aldehyde** is the $-C\overset{\displaystyle O}{\underset{\displaystyle H}{}}$, formyl group. The general formula is

RCHO, where R represents a hydrocarbon radical.

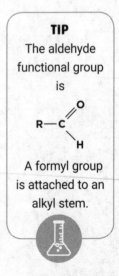

TIP

The aldehyde functional group is

$$R-C\overset{\nwarrow O}{\underset{\searrow H}{}}$$

A formyl group is attached to an alkyl stem.

Preparation from an Alcohol

Aldehydes can be prepared by the oxidation of an alcohol. This can be done by inserting a hot copper wire into the alcohol. A typical reaction is:

$$
\underset{\substack{\text{methanol} \\ \text{or} \\ \text{methyl alcohol}}}{\text{H}-\overset{\displaystyle\text{H}}{\underset{\displaystyle\text{H}}{\text{C}}}-\text{OH}} + [\text{O}] \longrightarrow \left[\text{H}-\overset{\displaystyle\text{H}}{\underset{\displaystyle\text{OH}}{\text{C}}}-\text{OH}\right] \longrightarrow \underset{\substack{\text{methanal} \\ \text{or} \\ \text{formaldehyde}}}{\text{H}-\overset{\displaystyle\text{O}}{\underset{\displaystyle\text{H}}{\text{C}}}} + \text{H}_2\text{O}
$$

The middle structure is an intermediate structure; since two hydroxyl groups do not stay attached to the same carbon, it changes to the aldehyde by a water molecule "breaking away."

The aldehyde name is derived from the alcohol name by dropping the -*ol* and adding -*al*.

Ethanol forms ethanal (acetaldehyde) in the same manner.

Organic Acids or Carboxylic Acids

The functional group of an organic acid is the $-\text{C}\overset{\displaystyle\text{O}}{\underset{\displaystyle\text{OH}}{}}$, carboxyl group. The general formula is R—COOH.

Preparation from an Aldehyde

Organic acids can be prepared by the mild oxidation of an aldehyde. The simplest acid is methanoic acid, which is present in ants, bees, and other insects. A typical reaction is:

$$
\underset{\substack{\text{methanal} \\ \text{or} \\ \text{formaldehyde}}}{\text{H}-\overset{\displaystyle\text{O}}{\underset{\displaystyle\text{H}}{\text{C}}}} + [\text{O}] \longrightarrow \underset{\substack{\text{methanoic acid} \\ \text{or} \\ \text{formic acid}}}{\text{H}-\overset{\displaystyle\text{O}}{\underset{\displaystyle\text{OH}}{\text{C}}}}
$$

Notice that in the IUPAC system, the name is derived from the alkane stem by adding -*oic*.

Ethanal can be oxidized to ethanoic acid:

$$
\underset{\text{ethanal}}{\text{H}-\overset{\displaystyle\text{H}}{\underset{\displaystyle\text{H}}{\text{C}}}-\text{C}\overset{\displaystyle\text{O}}{\underset{\displaystyle\text{H}}{}}} + [\text{O}] \longrightarrow \underset{\substack{\text{ethanoic acid} \\ \text{or} \\ \text{acetic acid}}}{\text{H}-\overset{\displaystyle\text{H}}{\underset{\displaystyle\text{H}}{\text{C}}}-\text{C}\overset{\displaystyle\text{O}}{\underset{\displaystyle\text{OH}}{}}}
$$

Acetic acid, as ethanoic acid is commonly called, is a mild acid that, in the concentrated form, is called glacial acetic acid. Glacial acetic acid is used in many industrial processes, such as making cellulose acetate. Vinegar is a 4% to 8% solution of acetic acid that can be made by fermenting alcohol.

It is possible to have more than one carboxyl group in a carboxylic acid. In the ethane derivative, it would be ethanedioic acid with a structure like this:

$$\underset{\text{ethanedioic acid}}{\text{H–O–}\overset{\displaystyle \overset{O}{\|}}{\text{C}}\text{–}\overset{\displaystyle \overset{O}{\|}}{\text{C}}\text{–O–H}}$$

Summary of Oxygen Derivatives

Functional Group

$$\underset{\substack{\text{hydrocarbon}}}{\text{R—H}} \rightarrow \underset{\substack{\text{chlorine}\\\text{substitution}\\\text{product}}}{\text{R—Cl}} \rightarrow \underset{\substack{\text{alcohol}\\\text{(ending -ol)}}}{\text{R—OH}} \rightarrow \underset{\substack{\text{aldehyde}\\\text{(ending -al)}}}{\text{R}^1\text{CHO}} \rightarrow \underset{\substack{\text{acid}\\\text{(ending -oic)}}}{\text{R}^1\text{—COOH}}$$

Note: R^1 indicates a hydrocarbon chain different from R by having one less carbon in the chain.

An actual example using ethane:

$$\underset{\text{ethane}}{C_2H_6} \xrightarrow{Cl_2} \underset{\text{chloroethane}}{C_2H_5Cl} \xrightarrow{NaOH} \underset{\text{ethanol}}{C_2H_5OH} \xrightarrow{[O]} \underset{\text{ethanal}}{CH_3CHO} \xrightarrow{[O]} \underset{\substack{\text{ethanoic}\\\text{acid}}}{CH_3COOH}$$

Ketones

When a secondary alcohol is slightly oxidized, it forms a compound having the functional group $\text{R—}\overset{\displaystyle \overset{\text{|}}{\underset{\underset{O}{\|}}{C}}\text{—R}^1$ and is called a ketone. The R^1 indicates that this group need not be the same as R. An example is:

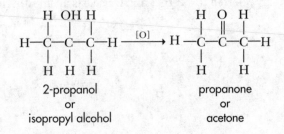

2-propanol
or
isopropyl alcohol

propanone
or
acetone

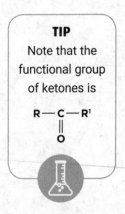

TIP

Note that the functional group of ketones is

$\text{R—}\overset{\displaystyle \overset{\|}{O}}{C}\text{—R}^1$

Example in a longer chain:

$$H-\overset{\overset{\displaystyle H}{|}}{\underset{\underset{\displaystyle H}{|}}{C}}-\overset{\overset{\displaystyle O}{\|}}{C}-\overset{\overset{\displaystyle H}{|}}{\underset{\underset{\displaystyle H}{|}}{C}}-\overset{\overset{\displaystyle H}{|}}{\underset{\underset{\displaystyle H}{|}}{C}}-H$$

butan-2-one

TIP

The functional group of ethers is R–O–R^1.

In the IUPAC method, the name of the ketone has the ending *-one* with a digit indicating the carbon that has the double-bonded oxygen preceding the ending in larger chains, as shown in butan-2-one. Another method of designating a ketone is to name the radicals on either side of the ketone structure and use the word **ketone**. In the preceding reaction, the product would be dimethyl ketone.

Note that both aldehydes and ketones contain the carbonyl group in their structures. In the aldehydes, it is at the end of the chain, and, in acids, it is the interior of the chain.

Ethers

When a primary alcohol, such as ethanol, is dehydrated with sulfuric acid, an **ether** forms. The functional group is R—O—R^1, in which R^1 may be the same hydrocarbon group, as shown in example 1 below, or a different hydrocarbon group, as shown in example 2.

1.

ethanol + ethanol $\xrightarrow{H_2SO_4}$ ethoxyethane (ethyl ether or diethyl ether) + H_2O

2. Another ether with unlike groups, R—O—R^1:

ethoxypropane (ethyl propyl ether)

In the IUPAC method, the ether name, as shown in the examples, is made up of two attached alkyl chains to the oxygen. The shorter of the two chains becomes the first part of the name, with the *–ane* suffix changed to *–oxy* and the name of the longer alkane chain as the suffix. Examples are ethoxyethane and ethoxypropane.

Diethyl ether is commonly referred to as ether and is used as an anesthetic.

Amines and Amino Acids

The group NH_2^- is found in the amide ion and the amino group. Under the proper conditions, the amide ion can replace a hydrogen in a hydrocarbon compound. The resulting compound is called an **amine**. Two examples are shown in Figure 13.12 below.

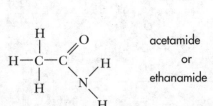

methylamine

aniline

FIGURE 13.12 Amine Compounds

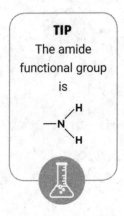

TIP

The amide functional group is

In *amides*, the NH_2^- group replaces a hydrogen in the carboxyl group. When naming amides, the *-ic* of the common name or the *-oic* of the IUPAC name of the parent acid is replaced by *-amide*. For example:

acetamide

or

ethanamide

Amino acids are organic acids that contain one or more amino groups. The simplest uncombined amino acid is glycine, or amino acetic acid, NH_2CH_2-COOH. More than 20 amino acids are known, about half of which are essential in the human diet because they are needed to make up the body proteins.

TIP

Esters can be compared to inorganic salts.

Esters

Esters are often compared to inorganic salts because their preparations are similar. To make a salt, you react the appropriate acid and base. To make an ester, you react the appropriate organic acid and alcohol. The process to make ethyl acetate is shown in Figure 13.13.

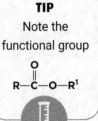

TIP

Note the functional group

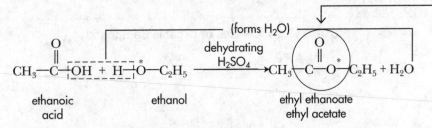

ethanoic acid

ethanol

ethyl ethanoate
ethyl acetate

FIGURE 13.13 Esterification

The name is made up of the alkyl substituent of the alcohol and the acid name, in which *-ic* is replaced with *-ate*.

The general equation is:

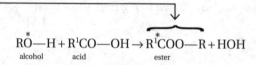

$$\overset{*}{R}O{-}H + R^1CO{-}OH \rightarrow R^1COO{-}\overset{*}{R} + HOH$$

alcohol   acid   ester

Esters usually have sweet smells and are used in perfumes and flavor extracts.

Figure 13.14 summarizes the organic structures and formulas discussed in this section.

Class	Functional Group	General Formula
Alcohol	$-$ OH	R $-$ OH
Alkyl halides	$-$ X X = F, Cl, Br, or I	R $-$ X
Ether	$-$ O $-$	R $-$ O $-$ R′
Aldehyde	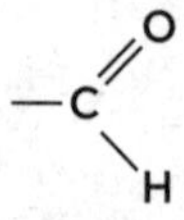 $-C\overset{\displaystyle O}{\underset{\displaystyle H}{}}$	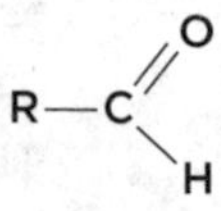 $R-C\overset{\displaystyle O}{\underset{\displaystyle H}{}}$
Ketone	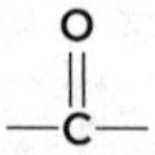 $-\overset{\displaystyle O}{\underset{}{C}}-$	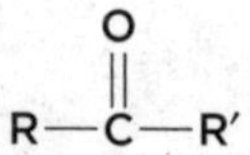 $R-\overset{\displaystyle O}{\underset{}{C}}-R'$
Carboxylic acid	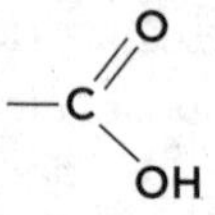 $-C\overset{\displaystyle O}{\underset{\displaystyle OH}{}}$	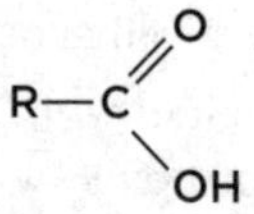 $R-C\overset{\displaystyle O}{\underset{\displaystyle OH}{}}$
Ester	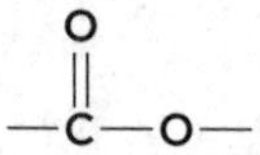 $-\overset{\displaystyle O}{\underset{}{C}}-O-$	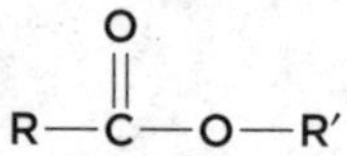 $R-\overset{\displaystyle O}{\underset{}{C}}-O-R'$
Amine	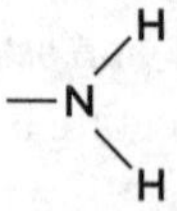 $-N\overset{\displaystyle H}{\underset{\displaystyle H}{}}$	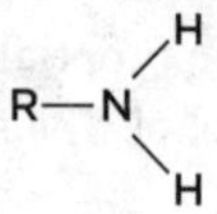 $R-N\overset{\displaystyle H}{\underset{\displaystyle H}{}}$

FIGURE 13.14 Classes of Organic Compounds

CHAPTER 14

The Laboratory

> **Learning Objectives**
>
> In this chapter, you will learn how to:
>
> o Name, identify, and explain proper laboratory rules and procedures.
> o Identify and explain the proper use of laboratory equipment.

Laboratory setups vary from school to school depending on whether the lab is equipped with macro- or microscale equipment. Microlabs use specialized equipment that allows lab work to be done on a much smaller scale. The basic principles are the same as when using full-sized equipment, but microscale equipment lowers the cost of materials, results in less waste, and poses less danger. The examples in this book are of macroscale experiments.

Along with learning to use microscale equipment, most labs require a student to learn how to use technological tools to assist in experiments. The most common are:

Gravimetric balance with direct readings to thousandths of a gram instead of a triple-beam balance

pH meters that give pH readings directly instead of using indicators

Spectrophotometer, which measures the percentage of light transmitted at specific frequencies so that the molarity of a sample can be determined without doing a titration

Computer-assisted labs that use probes to take readings, e.g., temperature and pressure, so that programs available for computers can print out a graph of the relationship of readings taken over time

The Ten Commandments of Lab Safety

The following is a summary of rules you should be well aware of in your own chemistry lab.

1. Dress appropriately for the lab. Wear safety goggles and a lab apron or coat. Tie back long hair. Do not wear open-toed shoes.
2. Know what safety equipment is available and how to use it. This includes the eyewash fountain, fire blanket, fire extinguisher, and emergency shower.

3. Know the dangers of the chemicals in use, and read labels carefully. Do not taste or sniff chemicals.

4. Dispose of chemicals according to instructions. Use designated disposal sites, and follow the rules. Never return unneeded chemicals to the original containers.

5. Always add acids and bases to water slowly to avoid splattering. This is especially important when using strong acids and bases that can generate significant heat, form steam, and splash out of the container.

6. Never point heating test tubes at yourself or others. Be aware of reactions that are occurring so that you can remove them from the heat if necessary before they "shoot" out of the test tube.

7. Do not pipette anything by mouth! Never use your mouth as a suction pump, not even at home with toxic or flammable liquids.

8. Use the fume hood when dealing with toxic fumes! If you can smell them, you are exposing yourself to a dose that can harm you.

9. Do not eat or drink in the lab! It is too easy to take in some dangerous substance accidentally.

10. Follow all directions. Never haphazardly mix chemicals. Pay attention to the order in which chemicals are to be added to each other, and do not deviate!

Some Basic Setups

Throughout this book, drawings of laboratory setups that serve specific needs have been presented. You should be familiar with the assembly and use of each of these setups.

The following are additional laboratory setups with which you should be familiar:

1. Preparation of a gaseous product, soluble in water and lighter than air, by the downward displacement of air. See figure 14.1.

EXAMPLE

Preparation of ammonia (NH_3).

$$2NH_4Cl(s) + Ca(OH)_2(s) \rightarrow CaCl_2(s) + 2H_2O(g) + 2NH_3(g)$$

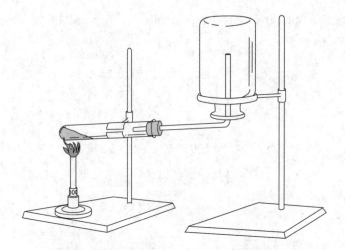

FIGURE 14.1 Preparation of Ammonia

2. Separation of a mixture by chromatography. See figure 14.2.

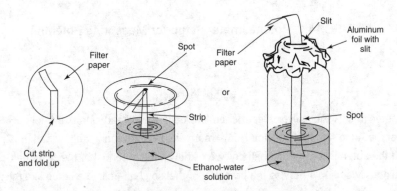

FIGURE 14.2 Chromatography Setup

EXAMPLE

Chromatography is a process used to separate parts of a mixture. The component parts separate as the solvent carrier moves past the spot of material to be separated by capillary action. Because of variations in solubility, attraction to the filter paper, and density, each fraction moves at a different rate. Once separation occurs, the fractions are either identified by color or removed for other tests. A usual example is the use of Shaeffer Skrip Ink No. 32, which separates into yellow, red, and blue streaks of dyes.

3. Measuring potentials in electrochemical cells. See figure 14.3.

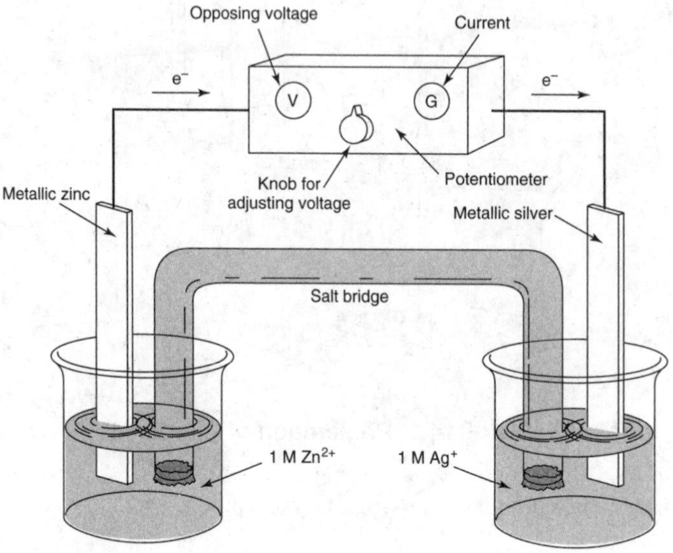

Opposing voltage

Current

e^- →

V   G

← e^-

Metallic zinc

Knob for
adjusting voltage

Potentiometer

Metallic silver

Salt bridge

1 M Zn²⁺   1 M Ag⁺

FIGURE 14.3 Potentiometer Setup for Measuring Potential

EXAMPLE

The voltmeter in this zinc-silver electrochemical cell would read approximately 1.56 V. This means that the Ag to Ag^+ half-cell has 1.56 V more electron-attracting ability than the Zn to Zn^{2+} half-cell. If the potential of the zinc half-cell were known, the potential of the silver half-cell could be determined by adding 1.56 V to the potential of the zinc half-cell. In a setup like this, only the difference in potential between two half-cells can be measured. Notice the use of the **salt bridge** instead of a porous barrier.

---

4. Replacement of hydrogen by a metal. See figure 14.4.

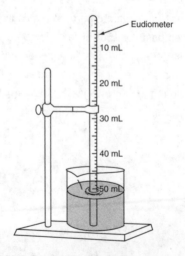

Eudiometer

10 mL

20 mL

30 mL

40 mL

50 mL

FIGURE 14.4 Eudiometer Apparatus

EXAMPLE

Measure the mass of a strip of magnesium with an analytical balance to the nearest 0.001 g. Using a coiled strip with a mass of about 0.040 g produces about 40 mL of H_2. Pour 5 mL of concentrated HCl into a eudiometer, and slowly fill the remainder with water. Try to minimize mixing. Lower the coil of Mg strip into the tube, invert it, and lower it to the bottom of the beaker. After the reaction is complete, you can measure the volume of the gas released and calculate the mass of hydrogen replaced by the magnesium.

Index